From the Soil Up

Donald L. Schriefer

Acres U.S.A.
Austin, Texas

FROM THE SOIL UP

Acres U.S.A., Publishers
P.O. Box 91299, Austin, Texas 78709 U.S.A.
phone (512) 892-4400 • fax (512) 892-4448

Printed in the United States of America

Publisher's Cataloging-in-Publication

Schriefer, Donald L
From the soil up: a systems approach to crop & soil management / Donald L. Schriefer — 1st ed.
p. cm.
Library of Congress Catalog Card Number: 99-76579
ISBN: 0-911311-63-7

1. Agricultural systems. 2. Soil science. 3. Organic farming. 4. Sustainable agriculture. I. Title
S494.5.S95S37 2000 631.4
QBI99-1665

ABOUT THE AUTHOR

Donald L. Schriefer passed from this life on July 30, 1998. He had spent more than five years battling acute leukemia, but he did not lie down and wait for death to come. He left this manuscript as a legacy to his lifelong friends — the farmers — knowing that those left behind would have it published.

Schriefer lived with his wife Susan in the small farming community of DeMotte, Indiana. His background included a B.S. from the University of Illinois and postgraduate work at the University of Wisconsin at River Falls. Early in his career he taught vocational agriculture to high school students and conducted one of the largest educational programs for farmers in Illinois.

More recently he incorporated Advanced Agriculture, Inc. as a vehicle for his Soils and Crops consulting business. Schriefer may be best known for the "systems" approach in his farm management programs.

Making education an integral part of his personal philosophy, Schriefer contributed frequently to various agricultural publications and was well known for conducting numerous seminars and farm programs annually. He has previously written two books, *Agriculture in Transition* and *Tillage in Transition.*

Schriefer's interests revolved around the improvement of farming practices as they related both to profits and the conservation of soil, nature and wildlife. With his extensive understanding of soils, Schriefer played a major role in the development of a primary tillage concept which became known as disc-chiseling.

Schriefer was a pioneer in the development of the concept of relating tillage to soil fertility based upon the impact of tillage on the three major "soil basics." Relating tillage to the management of soil aeration, soil water and residue decay has brought major changes in the designs of tillage tools and tillage practices. Some of his unique approaches to the business of farming include the "systems approach" and the development of the "yield-limiting factor" concept as a foundation for maximizing yields, quality and net returns. Very early in his career, Schriefer recognized the serious yield-limiting factors caused by the natural "pan" under most Midwestern soils.

Terms now commonplace in the agricultural community, such as "crop and soils weatherproofing," the "row support system," the "zone of updraft concept," and the "tillage commandments" were originally inspired by Schriefer. He understood the necessity of coordinating every production management practice with the purpose of capturing more sunlight and carbon dioxide per acre by increasing chlorophyll and photosynthesis. During the final years of his life, Schriefer was involved in promotion and testing the use of a foliar application program that, coupled with complete soil management, may salvage crops or take good crops to higher yield levels.

CONTENTS

FOREWORD

The business of farming may be compared to a road show that opens in the spring and closes in the late fall or early winter. The show portrays an intense interaction between the soil-plant marriage with the farmer serving as the director. In this capacity the farmer must annually renegotiate his contract, and if he is successful at his trade, he may renew it approximately thirty-five times during his career.

Farming demands the highest degree of professionalism on the part of its director. A successful farming operation is a professional program that will not tolerate long term tenure for amateur directors.

A successful farmer-director will have developed a working philosophy toward his business. He will know that agriculture has elevated mankind from a "hunt and gather" society to one of total dependence upon organized food production. As a result modern agriculture has freed people from the full-time need to hunt for food causing major urbanization of mankind at huge expense to the environment, resources and other forms of life. The responsibility for producing an abundance of high quality foods along with conservation of soil and the realization that other forms of life also have rights to the land falls directly onto the shoulders of the farmer.

A successful farmer will enjoy his profession while being keenly aware that he does not have as much freedom as is often attributed to him. By being self-employed he can arrange his daily schedule, decide when to buy and sell, determine his crop rotation, varieties to plant, and so on. In effect the farmer does have much control over his own destiny, but in other important areas relating to his welfare he has lost almost all control. For example he has very little political influence. He cannot regulate supply to demand and has lost additional strength by his failure to organize or properly align himself with the many agri-related businesses he so amply supports.

A thoughtful farmer will tire of living on hope or prospering only because others have failed. He will realize that his well-being must be tied to supply and demand while competing in a world marketplace. Sooner or later the entire agri-industry must organize to control the whole. Total industry-wide organization would not be a loss of freedom but rather an expression of the intelligent use of this freedom.

A successful farmer also realizes that he is not the producer just as a physician is not the healer. The plant, with its umbilical-like connection to the soil, is the true producer with the farmer playing a supportive role in the process.

In this supportive role there are many factors interacting within the soil-plant system that will determine a farmer's success or failure. Farming is not as big a gamble as many would believe since the production factors within the farmer's control are the ones most influential in determining his success.

Profitable production is dependent upon the removal or minimization of yield-limiting factors. These yield-limiting factors must first be identified by the farmer so that they can be removed. Because their identification demands an in-depth knowledge of the total soil-plant system the farmer must often revert to the status of "student" while he is practicing his profession.

His education must be orderly and sequential. Too often agricultural education is product orientated, disjointed and very confusing. My purpose in writing FROM THE SOIL UP is to provide a regular sequence of information to help you identify and remove crop limiting factors and also to help you completely systematize your entire farm production program.

As your education progresses and successes multiply you will no longer derive pleasure from simply outproducing your neighbor, but rather from serving as an example for others to emulate.

My challenge is for you to become the best farmer in your community and then to serve as a leader by graciously sharing your knowledge with others. Recognize your potential, develop it and pass it on to others. In addition to improving your own community it will give you a good feeling of satisfaction and accomplishment. *You* must become the "professional" by being capable of identifying and removing yield-limiting factors. That is what this book is all about.

BASIC SOILS

A study of the basic soil-forming processes is a prerequisite to good soil husbandry. Without this background we tend to view soils as perpetual and fail to consider the dire consequences of soil destructive practices.

To fully appreciate our conservation problems, consequences and ultimate solutions we must review the basics of the soil-forming processes. All soils are derived primarily from rocks. Rocks, however, are not soil and as such are not plant supportive. A large rock may be troublesome when struck with a tillage tool but it must be appreciated for what it really is.

A rock is made of a single mineral or a combination thereof. These minerals in turn contain elements that plants require, such as calcium, phospherous and potassium. In other words, the elements required by plants are bound up within the mineral chemistry which in turn is cemented into the rock, and as such the life-supporting elements are unavailable for plant usage.

Rocks, as illustrated in Figure 1, are simply plant nutrients and potential soil in storage. As a grain of wheat must be ground before it can be made into flour and bread, a rock also must be ground and reduced to "flour" before it can be formed into soil.

What forces then were responsible for the grinding or reducing of all of those rocks into "flour" and eventually into soil? The soil-forming processes responsible for our present soils have been taking place over a period of many thousands of years. The midwest soil story goes back ten to fifteen thousand years to a time when the area was last visited by a glacial intrusion known as the Wisconsin Glacier.

Glaciers are, to say the least, an awe-inspiring phenomena of nature. They are cyclic in the sense that they seem to appear about every ten to twenty thousand years. The earth's crust shows telltale evidence of at least four to five glacial visits in the past. There may have been many more.

It would be easy for us to view a glacial period as a natural disaster. It would certainly be disastrous for our present civilization if we were plunged into a glacial period. But we must view glaciers as part of a great plan that provides for continuing life on the planet Earth. Soils go through an aging process similar to life itself. A soil that is thirty to forty thousand years old, in humid regions, weathers and leaches to the point that it may no longer sustain the higher forms of seed-producing plants needed for animal nutrition. Therefore, we probably should view glacial periods as an essential renewal process, a springtime renewal of soil and life that occurs every twenty thousand years or so.

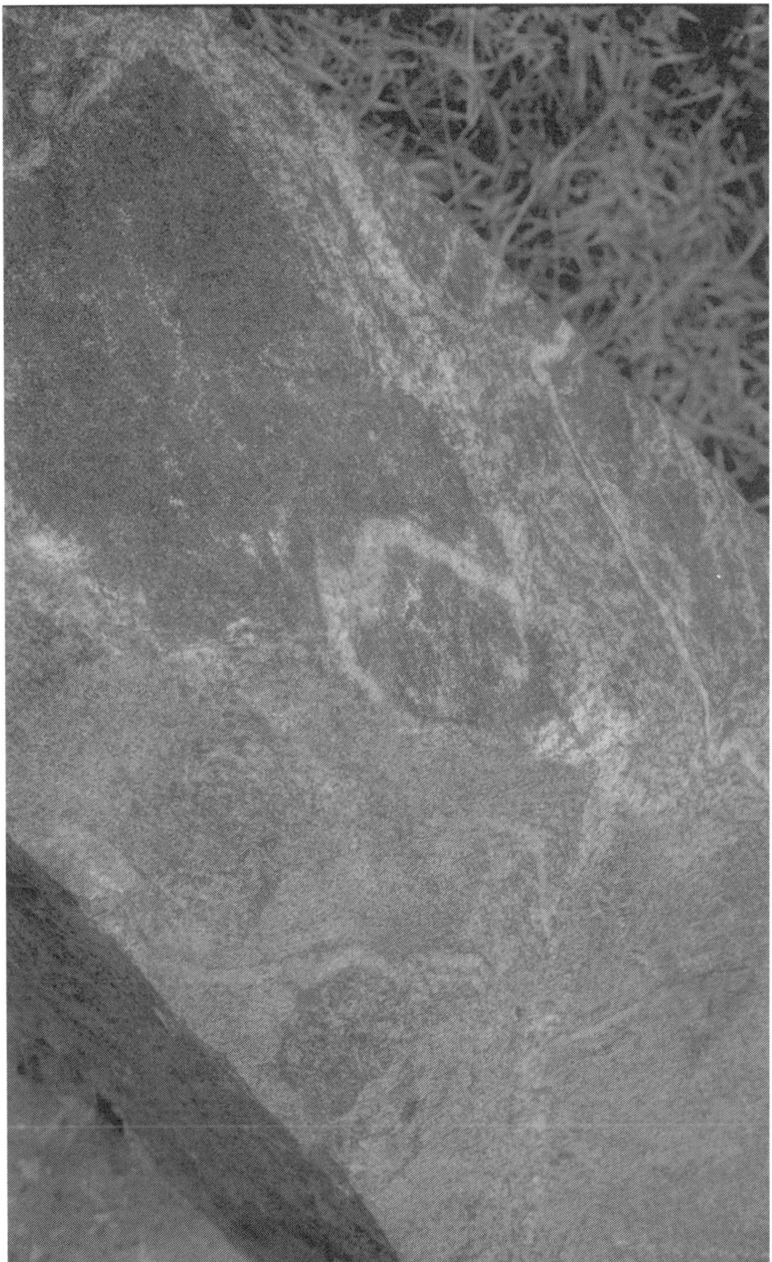

Fig. 1

Glacial periods were preceded by a few degrees of cooling of the earth's atmosphere. The theories of why the earth's atmosphere cools periodically range from "sunspot activity" and "volcanic activity" to a "periodic tilting of the earth by a few degrees." No one knows the reasons for certain, but many lean toward the "tilting earth" theory. Glacial periods are preceded by a buildup of the northern polar cap. When the ice cap builds to a given thickness the mass begins to flow outward because of rounded ball-bearing-like crystalization at the base of the pack, or the plasic flow characteristics of ice under great pressure. The outward flow continues as long as pressure is maintained at the polar region due to the continued snowfall.

The size of glaciers and their affect on the landscape can stagger one's imagination. The Wisconsin Glacier lay across Wisconsin and most of Illinois at a thickness that may have exceeded one mile. Some believe the last glacier must have been more than two miles in thickness for it to have gouged out the Great Lakes. Glaciers locked up so much water that the ocean levels dropped, thus exposing vast areas of land along their shores that are now again under water. Most of the Continental Shelves were probably exposed.

The bulldozing effect of the Wisconsin Glacier created the Great Lakes and caused soil compression that is still rebounding under the lakes and exposing shoreline on one side while creeping inland on the opposite shore. Many northern lakes were formed when huge blocks of ice broke away, silted in and melted, leaving holes filled with water. Streams, rivers, lakes, sand ridges, gravel deposits and rounded granite boulders which are hundreds of miles from their source in Canada are all evidence of glacial visits.

Glaciers act as huge bulldozers, rooting out the earth's crust which includes blocks of rocks. This material is mixed and ground up within the glacial mass much like feed in the gizzard of a bird. The rocks, similar to kernels of wheat, must be ground as the first step in preparing the recipe for soil.

The granite rock in Figure 1 shows the effects of its long ride within the glacier. At the beginning of its trip this rock was probably a square or rectangular mass and much larger than its present size. The "gizzard" action within the glacier has slowly ground most of its original mass into rock "flour" during the long ride from its Canadian source.

Eventually the forward motion of the glacier stopped and it slowly retreated by melting away. What remains is called parent material, or the stuff from which new soils are formed.

Rock flour, however, is still not soil. To become soil the fine mineral particles must go through a weathering process that releases the individual nutrient elements needed by plants. This weathering process is both physical and chemical.

The rate of soil formation and the kind of soils formed depends on the parent material left by the glacier and this material's interaction with its environment. Drainage, rainfall, slope, temperature and eventual vegetation are the major environmental factors that determine the kind of soils formed from parent material in any given area. With the vast array of parent material and the many microenvironmental conditions, we can begin to understand why there are so many different soil types, even within the confines of an individual farm.

As the glaciers retreated, soil life and the seeds of plants were waiting in the rubble for the chemical reactions within the presoil material to release the nutrient elements to support their needs. As soil life and plants multiplied the soil formed at an ever-increasing rate. Their life and death cycles slowly built a humus-laden topsoil environment that steadily improved to the benefit of both. Communites of life evolved. These communities included living organisms within the soil and communities of plants above the soil. As the soils developed nature proceeded to match plant species to soil variables until the sorting eventually terminated in an ideal marriage of plant species to soil conditions.

We use the term, "climax species" to describe the plant that is best adapted to a given set of soil and climatic conditions. In some areas hardwood and softwood trees predominate, while in others the prairie grasses climaxed and continued to adapt and improve. A climax species existed until climate or soil conditions changed and forced it to give way to a more worthy opponent.

The climax species had a great impact on the characteristics of present-day soils. The deep-rooted prairie grasses formed deep, black, humus-laden topsoils in some regions because of the annual underground death and recycling of hundreds, or even thousands, of pounds of roots per acre. By contrast, forest soils developed very shallow topsoils because of the long-lived roots and surface decomposition of leaves. Very little decay takes place below the ground in forests as compared to prairies.

In the ten thousand or more years since the last glacier receded nature has established an ecologically intricate soil, plant and animal balance. The exacting laws of nature rule the scene in an impartial and unyielding fashion. Violators of the natural laws have never failed to pay the penalties and the natural processes have brought about a continuous improvement through natural selection and recycling. Into this harmonious environment walked man with his technological agriculture.

Man's agriculture has two traits that are contrary to nature's way, but more or less necessary to his. First, agriculture has inserted plant species into areas that are far less adapted to them than the natural species they replace. This has necessitated the development of a host of cultural practices to assure their survival. Left unattended, the newly introduced crop species would fail. Second, in order to eliminate the natural species and to prepare a bed for precise seed placement, man had to introduce tillage practices that left the ground bare and exposed.

Nature's method was the ultimate in soil conservation and soil building while basic agriculture set the stage for potential destruction of the soil system. Keep in mind that the soil-plant system was designed to exist on it own independent of higher animals. Agriculture, however, is primarily based upon the support of people and animals.

Most crops are produced specifically to support animals. This introduces some very interesting and different concepts that can be in opposition to the natural system. Production is removed and transferred to feedlots. Then livestock and its byproducts leave the farm to support the needs of urban areas. Nutrients from the soil are transferred from the farm and eventually enter waterways through sewage disposal or else end up embalmed in burial vaults. This is contrary to the natural way of recycling. If continued this will surely limit man's proliferation and endurance as a species.

We understand the background of the soil-forming processes and nature's highly successful ways of producing while at the same time improving and conserving. Of equal importance is the need to recognize that agricultural technology is quite different from nature's proven ways and is basically soil destructive.

Our goals must be to implement those production practices that absolutely minimize damage to the soil system. This will require an in-depth understanding of the total soil system and implementation of technology that will complement and enhance this natural resource that makes everything else possible.

This leads to the subject of tillage, an area that may have the most important impact on production and soil longevity. We must realize that tillage is an unnatural process that may be necessary, but nevertheless is soil destructive, particularly because of erosion.

The welfare of our soils, their ability to produce and their longevity will be determined mostly through tillage methods. Our present tillage practices are too often proving not only to be destructive of the soil, but also yield limiting. Many claim it to be a damning indictment against American technology, research and education that we permit such soil-wasting practices to continue. Our intent, however, is not to accuse at random, but rather to define the problems and then to explore intelligent solutions.

A major abuse of our most productive soils is caused by the intensity in which they are farmed. We tend to avoid the inconveniences of conservation practices. We have been lulled into the belief that Class I level land is free from soil losses through erosion and we are now paying the price. On the other hand, our most erosion-prone and less productive soils demand and often receive better conservation management such as meadow rotations, strip farming and contouring.

Erosion is so severe and environmentally troublesome that our society is beginning to question whether the rights of ownership of land also give the right to destroy it, as illustrated in Figure 2.

Fig. 2

We can rest assured that if we, as farmers, do not take the initiative in the conservation of our soils we will be forced to do so through legislation. This would be the ultimate insult to the entire farming industry and it need not come about because today we have available the knowledge and the tillage systems which, properly implemented, will slow soil losses to acceptable levels.

Progress in soil conservation is dependent upon our acknowledgment of the extremely fragile and thin mantle of soil that sustains all of humanity. Everything of importance to our physical existence comes "from the soil up."

As far as we are concerned, we have the use of the only soil there is, although if destroyed by us it will eventually be renewed by some future glacial cycle. That may be too late for our species because we will most certainly degenerate together.

With these things in mind, let us get down to the business of understanding the requirements and the relationship between the soil-plant system. The soil-plant relationship is what the business of farming is all about. How well we understand this system will determine our worth as farmers. This generation must hold and operate the land in trust and pass it on to the next generation in better stead than received by us. To this extent, the knowledge and experience of past generations must become an ever-growing pool of knowledge. Each generation has the privilege and duty to share and add to this knowledge.

THE SOIL-PLANT SYSTEM

The green plant is truly a pinnacle of creation. It is the food and energy machine of the world. Nearly every living creature depends directly, or indirectly, on the green plant for nutrient needs.

The green plant is the "energy sink" for the earth. Without the conversion of the energy of the sun into carbohydrates through photosythesis this incoming energy would be a fleeting visitor and quickly lost to space by radiation. We would have no energy reserves from the past such as coal and petroleum. In fact, without the green plant serving as an oxygen generator and convertor of sunlight into usable energy forms the earth would be as lifeless as its neighbors.

The green plant may be considered the factory and we who are involved in its culture as plant, or factory, managers.

The ability of a plant to produce food begins with the process of photosynthesis, a complex process, as illustrated in Figure 3, of combining the energy of the sun with carbon dioxide and water into a simple sugar leaving oxygen as the byproduct.

$$\left.\begin{array}{c}\text{SUNLIGHT}\\ \&\\ \text{CHLOROPLAST}\end{array}\right\rangle + 6CO_2\ \&\ 12H_2O \rightarrow \underset{(C_6H_{12}O_6)}{\text{SUGAR}} + 6H_2O\ \&\ 6O_2$$

Fig. 3

This entire process takes place within the individual chlorophyll molecule known as the chloroplast. The process converts a fleeting energy source into a durable carbohydrate. This simple carbohydrate serves as the battery, or energy source, for all other creatures, both micro and macro. The earth's entire bio-mass exists because of this original energy source.

The plant uses the external energy of the sun to produce glucose sugar, which then becomes its internal energy source. Utilizing this internal energy, the plant takes up the necessary nutrients from the soil reservoir and proceeds to develop a genetically controlled nutrient package that is unique and different for each species of plant.

The many different plant species, each producing a different nutrient combination, provide a smorgasbord of nutrients that singularly or in combinations supply the nutrient needs of every form of non-plant life on earth. Although many forms of life secure their nutrients indirectly it may be assumed that most of the nutrients in non-plant life originated from the plant source.

In modern agriculture we are concerned with only a very small percentage of the total plant species on our planet. These are the crop species that have been specifically modified to meet our many needs. How well these selected species perform and meet our needs will depend on the depth of our understanding and our ability to manage the soil-plant system.

This becomes even more significant as we realize that most crop species are not nearly as hardy or as well adapted to the areas where they are planted as the natural species. These well-adapted natural species are usually considered to be weeds by the farmer who wages continuous war to prevent them from obliterating the less adaptable crop species.

A host of cultural practices had to be developed to assure crop survival. Agriculture involves the implementation of cultural practices which range from chemical to physical manipulation of the soil environment. Our objective is maximum performance in both quantity and quality of production.

Much knowledge and experience is necessary on the part of the farmer, or plant manager, to meet these objectives. This knowledge must begin with a general understanding of how the plant functions in relationship to its soil environment. A good basic understanding of this relationship must form the foundation around which the farming program is built. Because farming is totally concerned with the soil-plant system, a farmer's efforts relate to the management of this soil-plant system and his success will be in proportion to the degree in which his activities complement it.

A good beginning is the recognition that most of our production is given to us freely through the process of photosynthesis. For instance, if we were to burn one hundred pounds of dry crop material, including roots, leaves, stalks, and grain or fruit, we would discover that about ninety-five pounds of the entire material would vanish during the burning process. Approximately ninety-five percent of the entire plant is composed of sugars, starches, oils, cellulose, lignin and waxes that contain carbon, hydrogen and oxygen combinations. These originate from the original simple glucose sugar that was manufactured through the process of photosynthesis.

The original energy of the sun, plus the carbon dioxide and water that went into the manufacturing of these organic compounds, are all released in the process of burning. This release can take place through rapid combustion by fire or through the slow burn of metabolism within the cells of living organisms. Either process releases back into the environment the original energy from the sun along with the carbon dioxide and water that were originally combined through the process of photosynthesis.

After burning is completed, the five percent of material remaining is ash, or the mineral portion of the plant tissue which came from the soil reservoir. The five percent of ash considered here is based upon the assumption that the plant was grown in a fertile, well-balanced soil. Under adverse soil conditions the end result could be less than five percent of mineral.

The mineral and other nutrient contents of a crop cannot be guaranteed. The nutrient contents of a plant may vary greatly from what is considered normal just as quantity, or yields, can vary. These differences mainly reflect the plant's relationship to its soil environment and, to a lessor degree, its varietal differences. The "closer" it is to the soil, the greater the differences can be. In other words, the plant itself will vary more than its end products, the grain. Grain can also vary in its nutrient levels for a given variety, but these differences are usually not as pronounced as in the plant itself.

Assuming "average" nutrient values for any crop such as haylage, corn silage, celery, lettuce, etc. may be hazardous to the consumer's health whether it be man or beast. Also, severe excesses of a given nutrient within grains or forages can be just as detrimental as severe deficiencies. Our goal in the production of food must be to grow crops that are as genetically normal as possible. This means simply that we must do everything possible to avoid both excesses and deficiencies.

While no one really knows the perfect genetic balance for each crop species or variety, we believe if this balance could be met we would have met the ultimate goal in both quantity and quality.

Management of the ash content of crops is very important. In this respect we have two areas to consider. First, we must be certain the soil nutrients are in balance. Second, the soil system must be managed in a manner that will assure the recovery of the soil nutrients by the plant.

Soil balance and plant recovery are the primary purposes of soil management efforts. A perfect soil chemical balance is meaningless if the soil environmental conditions prevent the nutrient recovery, or uptake, by the roots. Plant nutrition involves much more than the simple act of applying fertilizers.

Another aspect to consider in the management of the nutrient needs of crops is that plants require both inorganic and organic nutrients or substances from the soil reservoir. The inorganic plant food can be purchased and is usually referred to as "fertilizer." The organic needs of the plant are real and extremely complex. This is the part of plant nutrition that cannot be purchased, but must come from the biological processes within a living soil.

It is interesting to reflect upon the fact that we may classify all living things as either producers or consumers. Plants are the only true producers, therefore providing either directly or indirectly the nutrient needs of the consuming world.

Consumers can be also divided into either above or below the ground categories. We tend to think of the above ground consumers as being large, or macro, in size and those underground as being mostly small, or micro, in size. In reality the micro and macro exist both above and below the ground.

The primary function of the consumers is to obtain their bodily needs by recycling other dead or living biomass. We often err by assuming that the large organisms are the most important of the consumers. The invisible micro-organisms within the soil mass are the real heroes of the consumer world. These are the organisms that have evolved into a symbiotic relationship with the plant world. They are life-supportive and not life-threatening. They consume and recycle dead plants, therefore supporting continuous life, as opposed to the above ground consumers who depend to a great extent on the consumption of living plants.

For all practical purposes the soil-plant system is a closed system, or circuit, and can function quite well without the above ground predators. It is good for humanity to be aware of this and recognize ourselves for exactly what we are — predators and not producers!

Soil life can manage the entire consumption of plant remains and maintain an endless balance of life and death without assistance from the larger, aboveground consumers. The myriad of underground micro-consumers function basically the same as the larger, above ground ones. The plant consumes carbon dioxide and gives off oxygen in producing sugar, while consumers must reverse the process by taking in large amounts of oxygen and giving off carbon dioxide to convert the sugar back into usable energy.

The balance between atmospheric carbon dioxide and oxygen is important and can be regulated entirely by the soil-plant system. This is illustrated in Figure 4.

Plants are designed to function in a symbiotic relationship with many species of soil organisms. The plant depends on the recycling process of soil decomposers. There is a symbiotic relationship between plants and the soil organisms that thrive along the rhizosphere, or root surfaces. Soil life is an integral part of the soil-plant system with the sole purpose of producing and recycling. Production could not continue very long without recycling.

The recycling process produces new combinations of compounds or nutrients not necessarily found in the original plants. Decomposition produces enzymes, hormones, antibiotics and organically

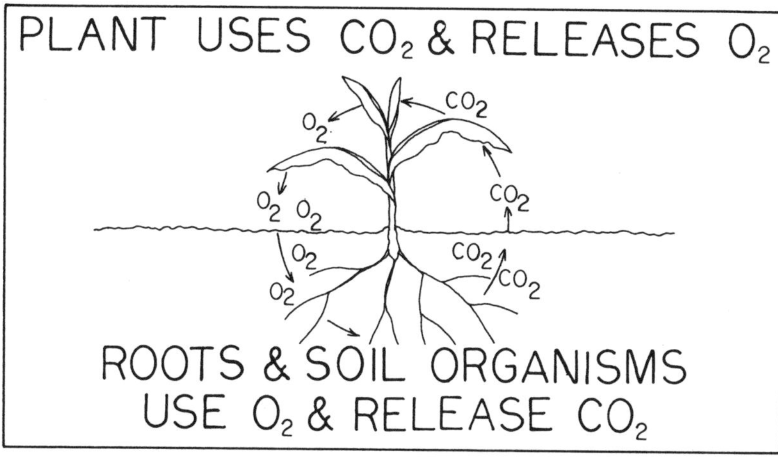

Fig. 4

chelated elements and possibly a host of yet undiscovered factors. These compounds are taken up and utilized by the living plant. This is the organic side of plant nutrition. It is the part of plant nutrition that cannot be purchased and must come via the biological processes of recycling.

The inorganic side of plant nutrition is much simpler than the organic. Plants use inorganic elements such as potassium (K), or compounds such as sulphates (SO_4) and nitrates (NO_3). The fertilizer industry deals with inorganics. Because the organics are biological products they are more complex and varied than inorganic elements or compounds.

The energy levels of a plant, its health, vitality and ability to produce are functions of the nutrient balance within the plant. Nutrient balance is more complicated than merely buying and spreading fertilizers. Crop species, in common with the wild plants, require both inorganic and organic nutrients. This is their nature and if we want excellent quantity and quality production we will consider both in our soil management practices because we will recognize that plant chemistry or nutrient content can vary greatly depending upon the ability of the soil to serve up nutrients. Since animal nutrition comes basically from the soil via the plant it is clear that deficient soils that produce deficient crops will in turn produce deficient animals. This is why the innocent assumption that each plant contains certain nutrient values may be hazardous to our health.

It is also dangerous to believe that technology evolved by man has all the answers. The truth is, we are still far from understanding the total soil-plant relationship. It is to our advantage, however, to do everything possible to benefit soil life because it improves plant performance. Recognizing the benefits is more important than understanding all of the reasons for it.

Approximately ninety-five percent of crop production comes to us free of charge. For instance, if we burned one-hundred pounds of dry, mature corn including roots, plant and ears, about ninety-five pounds of material would be lost in the burning process leaving about five pounds of ashes. The ninety-five pounds of material returned from where it came. The heat from the fire is the original energy from the sun that was captured by the chlorophyll. The carbon dioxide and water released during the burning process is the carbon dioxide and water that was synthesized into sugar during photosynthesis. About ninety-five percent of plants are made up of carbon, hydrogen and oxygen. It is this fraction that burns. The simple sugar containing only carbon, hydrogen and oxygen formed during photosynthesis is transformed into more complex carbon, hydrogen and oxygen compounds such as starch, oils, cellulose, waxes and lignin. These are the compounds that burn and represent the ninety-five percent of our production that comes to us "free" through the process of photosynthesis.

As plant managers, we manage only five percent of our total production. This five percent of mineral matter coming from the soil reservoir determines how efficient the plant factory will be. Improving the efficiency of the plant factory is our primary goal and a little improvement in the area of the five percent can greatly improve overall production.

The management of the soil-plant system is not an exact science because we do not thoroughly understand how it functions. A general understanding of the basic concepts relating to the soil-plant system may be achieved by comparing the system to an automobile's battery. See figure 5. In this comparison we can immediately see that the charging of the battery must keep pace with the discharge or the battery will fail. The charging and discharging of the battery is comparable to the factors involved with soil building and soil depletion.

Plant growth is necessary in the soil forming (charging) processes. In this respect the exterior energy coming from the sun reacts with carbon dioxide and water in photosynthesis making plant growth possible. Plants provide the organic residue materials that make it possible for the biological processes to function. The endless growth and biological recycling of organic materials within the soil provides the chemistry essential for mineral weathering and on-going soil forming processes.

Soil formation, beginning with the glaciers, is an ongoing process that will continue indefinitely. Biological activities are essential to this

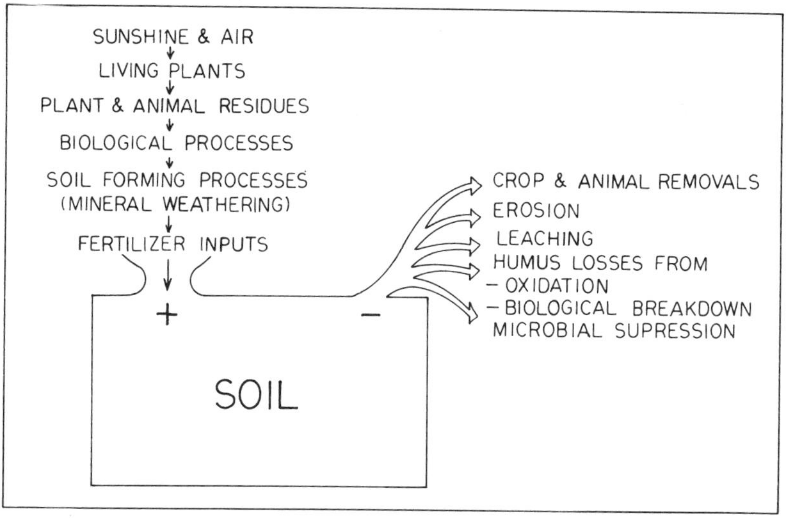

Fig. 5

overall process. There can be no true soils without the organic input.

Other soil charging factors are the fertility inputs that are now a standard practice in agriculture. Soil charging includes both organic and inorganic inputs.

The soil depleting, or discharging, factors include the removal of soil nutrients that are in crops and animal products which are sold off the farm, erosion, nutrient leaching, microbiological suppression and the depletion of humus due to oxidation, biological breakdown and, under certain conditions, liquification and leaching downward.

Obviously, the goal in agriculture is to maximize nutrient removal with crops and livestock and to absolutely minimize the "discharge" from erosion, leaching and humus loss. Losses of nutrients by removals through production can be easily replaced. Humus can even be rebuilt. Losses from erosion are similar to losses of lead from battery plates; it is permanent and irreversible.

We must not believe that massive nutrient removals through high production are destructive of the soil. On the contrary, high production should build soils. If high production does destroy soils it is due to faulty cultural practices which were used to obtain the high production. Understanding the battery analogy is important in laying the foundation for a successful farming operation.

Under natural conditions soil "charging" equaled or even exceeded soil "discharging" losses. Plant remains were simply recycled and humus continued to build. Erosion and leaching were minimal.

Modern agriculture, however, has drastically altered the situation. As farmers we must be aware of this and accept the responsibility of being soil stewards who preserve and protect the land and not believe, as many do, that soil destruction is a necessary part of modern farming. The farmer who can complacently watch his soils discharge through erosion is in the wrong business!

If as farmers we view ourselves as "plant managers" we can consider our role to be similar to that of a physician. Both must play supportive roles because the plant manager does not cause plants to produce, nor does the physician cause the healing. Both must learn to properly support life processes and to recognize their limitations. As the physician must have schooling, so must the farmer be well-schooled in the basic requirements of the soil-plant system.

In the process of managing the soil-plant system we must understand the basic needs of the plant in relationship to its soil environment.

As we work with and study the soil-plant system we become aware that there are certain basic needs that must be met if the system is to function properly. We believe there are four major soil basics that govern the welfare of the soil-plant system and are controllable by the farmer. These soil basics which must be managed by the farmer listed in their relative order of importance are, (1) **soil aeration,** (2) **soil water,** (3) **residue decay** and, (4) **soil fertility.** These basics must become the cornerstones of each farming operation. If we mismanage any one of these basics our farming operation will suffer accordingly.

To be meaningful our study of the soil-plant system must begin with the four soil basics and build from there. These soil basics are constant and represent nature's proven technology. Everything must be done to complement them for a farming operation to be successful.

With the understanding and correct management of the four soil basics we will have fulfilled our supportive role as farmers and nature will do the rest. Our soils will remain fully charged and production will far exceed that which has been experienced to date.

Increasing crop yields and quality is simply a matter of identifying and removing yield-limiting factors. We will find that most of our yield-limiting factors will be in association with the four soil basics — soil aeration, soil water, residue decay and soil fertility.

Let us now begin the process of identification and removal of yield-limiting factors with a thorough study of the four soil basics which provide the essential cornerstones for building our storehouse of knowledge.

SOIL BASICS

BASIC NUMBER ONE — MANAGING SOIL AIR

> We should be as attentive to suppying proper soil aeration as we are to assuring an adequate air supply to the cylinders of our tractor engines. Adequate soil aeration sets the stage for maximum production. Failure to manage soil air will guarantee crop failure even though everything else is done correctly.

The single, most important task of farmers is the management of soil air. Every life process is directly dependent upon the gasses oxygen and carbon dioxide. These are the two gasses with which we are concerned when we use the term, "soil air." Without carbon dioxide there could be no photosynthesis and therefore no food production. Without oxygen there could be no metabolization of the products of photosynthesis into usable energy.

To understand soil air, we will compare the soil and its atmosphere to an animal and its atmosphere. Both are dependent upon oxygen, both release carbon dioxide and both must exchange air through a process called, breathing. Each directly relies upon the proper exchange and balance of the gasses, oxygen and carbon dioxide.

The content of these gasses within the soil air is quite different from that which is found in the atmosphere above the soil. Soil air will almost always contain higher levels of carbon dioxide and lower levels of oxygen than is found in the surrounding atmosphere. The atmosphere contains about twenty-one percent oxygen while soil air rarely exceeds fifteen percent. The carbon dioxide levels can differ even more. The atmosphere contains about twenty-six hundredths of a percent of carbon dioxide, while the soil air may contain two to four percent. We see a similarity in animals. The air exhaled is much higher in carbon dioxide and lower in oxygen than the air inhaled.

There are logical reasons for these differences. In effect the soil is a complex blend of living systems that gives meaning to the often-used term, "the living soil." The living soil contains many forms of life, including the roots of plants and a complex blend of macro- and microbiological systems whose life functions consume oxygen and release carbon dioxide. In this respect the life processes within the soil are no different from that of a breathing animal. The intake of oxygen and the expulsion of carbon dioxide are necessary for both.

The major differences between soil life and animal life are the rate and method of air exchange. The carbon dioxide within an animal's blood triggers the desire to breathe and the animal will speed up or slow down the process according to its needs. Soils are not so controlled. The exchange of soil air is less precise and more dependent upon the physical factors influencing the inert and non-living soil mass.

It is necessary to understand how soils breathe because it can influence our decisions on tillage practices. Our tillage practices will have a major impact on how well the soils breathe. Breathing simply consists of the exchanging of air and this process within the soil is accomplished in four ways, with temperature, barometric pressure, rainfall and gas diffusion, each playing a major role.

TEMPERATURE

Soils absorb heat from the sun that, in turn, heats and expands the soil air, forcing part of it out into the atmosphere. When the sun goes down the soil cools, creating a partial vacuum which pulls air back into the pores of the soil. Through this process soils slowly inhale and exhale once every twenty-four hours. The degree to which a soil heats depends on its ability to absorb. A dark soil absorbs more heat than a light-colored soil because of differences in reflectivity. The desert nomads wore white clothing that reflected the sun's rays so they could remain cooler underneath. Excessive residue on the surface of the soil can also act as reflectors to slow the soil-warming process.

The amount of water within the soil also influences its rate of warmup. Water requires more energy to raise its temperature to a given level than does an equal volume of soil. When soils are holding high levels of moisture it will require more heat absorption to raise the temperature of the soil air to expand it outward. Drainage and water management become important factors in the rate at which a soil exchanges its air through the process of heat expansion.

BAROMETRIC PRESSURE

Barometric pressure changes cause a push and pull effect on soil air that brings about an air exchange. Moist air is lighter per volume than dry air. As moisture-laden air moves into an area, the barometer falls because the air pressure, or air weight, is decreased and exerts less push on the mercury. This lighter air would create a partial vacuum if the soil air was cooler than the atmospheric air. The soil air would then be pulled outward, thus creating an exchange through expansion. A cold, heavy air mass would have the opposite effect and tend to push air into the soil.

RAINFALL

Rainfall basically affects the soil air in two ways. As water moves downward through the soil it pushes the air out or down and draws new air in behind it. You can usually feel air coming out of a discharging field tile. This can result from soil air being pushed downward by insoaking water. Subsoil drainage and the soil's ability to take in water become important factors in soil air exchange.

Rain also absorbs a load of oxygen as it falls through the atmosphere and then carries it into the soil. A major portion of this oxygen which has been absorbed by rainfall can be used by the roots and soil organisms. An important source of oxygen is lost when water is permitted to run off the land instead of soaking into the soil. Rain water is better for the watering of plants than tap water because it contains more oxygen.

GAS DIFFUSION

The process known as "gas diffusion" is a very important method of soil air exchange. Diffusion is an equalization process that is a characteristic of all gasses. A law of gasses states that a gas will diffuse from an area of high concentration into an area of lower concentration. A high concentration of a particular gas, such as carbon dioxide, will move into an area of lower concentration in its effort to reach an equilibrium of pressure and concentration. It is good that gasses have this characteristic. Since carbon dioxide concentration is much higher in the soil air than in the atmosphere above ground it tends to diffuse out of the soil. Because oxygen concentrations are higher aboveground than within the soil air they will diffuse from the atmosphere into the soil.

The diffusion characteristic of gasses forces soils to breathe. We can immediately see that if this ability of gasses is to express itself soil conditions must permit the air to freely enter and exit. We can readily see what detrimental effect a sealing soil crust would have on this process.

Gas diffusion in the soil must be understood because it can be controlled to a large extent by tillage methods making it one process that is under the control of the farmer.

The soil's breathing process can be helped by the activities of soil organisms also. Living organisms provide their own built-in ventilation system through a soil granulation process which assures the exchange of air necessary for their survival. This is a symbiotic function which fulfills their need for air while also supplying air to the roots of the plants.

The beautifully granulated soil structure found under roadside sod is the result of well-established microbiological populations around the grass roots.

A living soil is a breathing soil. A biologically inactive, heavier-textured soil will breathe as slowly as an animal nearing death. In an animal there is no life without breath and in a soil there is no breath without life.

Soils must breathe simply to get carbon dioxide out of the soil and to get oxygen into the soil. This implies that soil air increases in carbon dioxide and decreases in oxygen, and in fact it does, because all of the living processes within the soil system consume oxygen and release carbon dioxide. This is similar to an animal's body which increases carbon dioxide and consumes oxygen through the process of metabolism. If oxygen cannot be replenished carbon dioxide builds at the expense of oxygen until life ceases.

IMPORTANCE OF CARBON DIOXIDE

The release of carbon dioxide into the soil by roots and soil microorganisms plays a major role in helping to reach our yield goals because it is absorbed into the leaf's chlorophyll and converted into sugar in the process of photosynthesis. Since the atmosphere is very low in carbon dioxide the soil source becomes an important factor toward increasing yields. If everything else is sufficient low carbon dioxide is the factor that can keep us from reaching maximum production.

To illustrate this point we will review one particular method of reaching extremely high yields. First a site must be selected in a low-lying area that is very high in organic matter. The soil must have good texture for aeration and adequate water and fertility.

The key is to get the site to release prodigious amounts of carbon dioxide. To do this, much of the carbonaceous organic matter or humus must be biologically consumed. This can be accomplished by using heavy rates of nitrogen which will build and sustain very high levels of microbial activity. The microbes, in turn, go into the organic matter for carbon energy and release large amounts of carbon dioxide as a product of their body respiration. This carbon dioxide diffuses out of the soil and temporarily tends to settle in the low-lying area. The corn leaves are bathed in an atmosphere rich in carbon dioxide and this greatly increases photosynthesis and sugar production leading to high yields.

You will note, however, that very rarely are record yields achieved more than once on the same site. This is because the first effort will decompose and use up so much of the original carbonaceous organic matter that there is not enough left over for an encore. The point to be made is that the soil must be a source of carbon dioxide for photosynthesis. It becomes obvious that a biologically inactive soil will not release the extra carbon dioxide and this partly explains lower yields on such soils.

The buildup of carbon dioxide within the soil system can be detrimental to the soil system. It must escape because as carbon dioxide levels build within, the soil oxygen must decline. A soil buildup of carbon dioxide simply means the soil has quit breathing for one reason or another. This can result in both an oxygen deficiency and carbon dioxide. toxicity. Carbon dioxide can combine with water, forming carbonic acid which can become toxic to roots and soil life. Oxygen starvation in combination with carbon dioxide toxicity is extremely detrimental to production.

IMPORTANCE OF SOIL OXYGEN

Oxygen is required for root extension, nutrient uptake and biological activity because all three require the release of energy in a manner similar to that released by an animal's exercising.

Figure 6 shows a very limited root system due to compaction which limits oxygen, while Figure 7 shows an extensive root system in the same soil type, but free of compaction. The difference is in the availability of oxygen which is absolutely essential for the extension of roots.

Roots are nomadic wanderers and, in the same manner as cattle crossing hilly terrain, will always take the path of least resistance. The

Fig. 6

Fig. 7

path of least resistance is always where there is enough oxygen. A root could grow through a very dense soil if enough oxygen could be fed to it there.

Wheel tracks imparted during planting and cultivation cause areas of extreme compaction. Rarely will a root be found in these compacted zones because soil compression pushes soil particles together and displaces soil air, as illustrated in Figure 8. The only

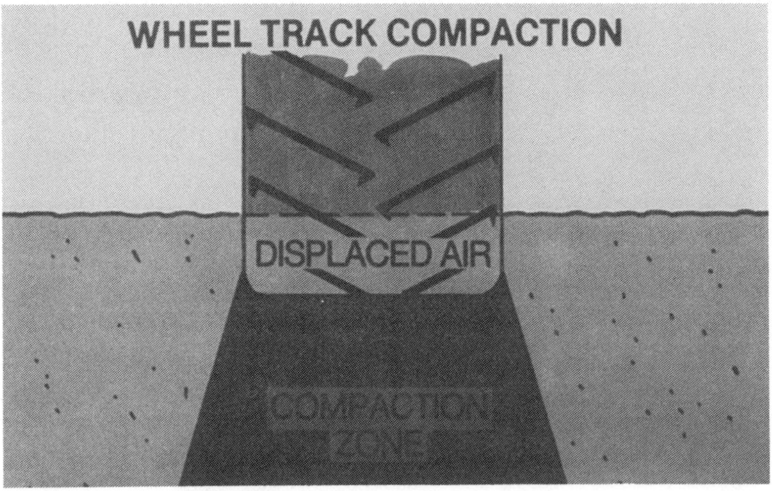

Fig. 8

roots found crossing a wheel track will be the ones which find a fracture line that contains some air. This is shown in Figure 6 where a lone, coarse-textured root is traversing the compaction zone.

Figure 9 is also a good example of a root growing through a fracture zone where it could find some oxygen. Note the herringbone growth pattern. The root is following the fracture zone and is incapable of penetrating the dense, airless clod itself. The right side of the picture illustrates the rapid proliferation of a root where it made contact with a properly formulated fertilizer band. This kind of proliferation is triggered by phosphorous uptake. However, phosphorous uptake and subsequent rapid root growth would not be possible without an adequate oxygen supply.

Fig. 9

In an area of soil that contains sufficient moisture, but no roots, you can be certain that there was not enough oxygen to support their growth. Understanding this will alter our methods of operation. Equipment should be sized to keep the wheel tracks, or compaction tracks, in the same place on all trips which follow primary tillage. A good practice would be to place markers on all pre-plant tillage equipment to make tracks stay in the same place.

The fertility in wheel tracks is not available if there are no roots there. Therefore, these zones will not contribute to production and must be minimized by keeping all of the tracks in the same place. Many farmers still sidedress nitrogen in the wheel tracks. What a waste it is to place fertilizer in areas that cannot be penetrated by roots! Wandering wheel tracks, as illustrated in Figure 10, must be avoided. No matter how well the surface is finished off, the deep, wedged compaction shown in Figure 8 is devastating to the roots.

Fig. 10

We know the importance of fertility placement as it relates to soil air, root growth patterns, and its subsequent uptake by the roots. After all, a crop's response to soil fertility is dependent upon the presence of roots and the uptake of the nutrients and both of these are oxygen-demanding functions. High soil fertility is meaningless without enough soil oxygen to support its uptake by the roots. Nutrient recovery is a major problem on many heavy-textured, but otherwise fertile, soils.

Root growth, nutrient uptake and biological activity function proportionately to the level of available oxygen within the soil system. Figure 11 graphically illustrates this relationship. The purpose of this graph is merely to illustrate a principle and does not result from exacting research data.

The exact level of oxygen within the soil air that is required for optimal growth, nutrient uptake and biological functions has not, to our knowledge, been positively established through research. We will assume that soil air contains a maximum of fifteen percent of oxygen by volume. Figure 11 shows that, as the percent of oxygen declines, the three important functions of nutrient uptake, root growth and biological activity will also decline. Intensively farmed soils of medium or heavy textures rarely contain optimal levels of oxygen to support maximum activity of these three functions.

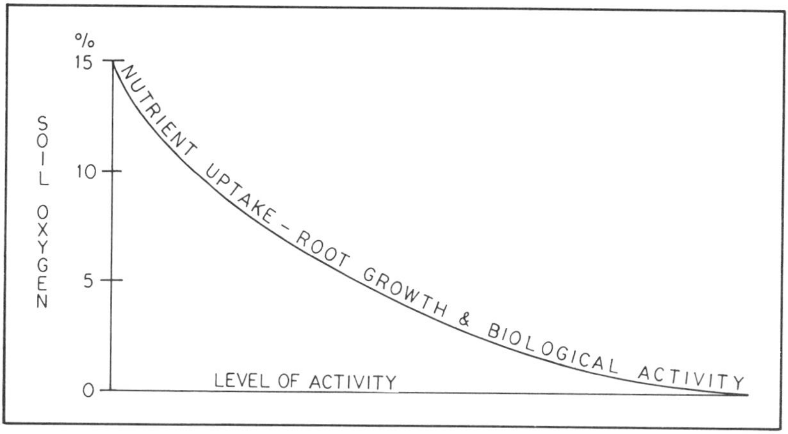

Fig. 11

Many factors must interact to get maximum production. A deficiency of soil oxygen creates a multitude of other deficiencies. Figure 12 is a familiar sight to every farmer. The corn in this area is showing deficiencies in several elements even though soil fertility levels of each element are high. The excessive water has displaced soil air. Without adequate oxygen the roots are incapable of taking in the water or nutrients that are in abundant supply and, as a result, the plants are starving to death.

Fig. 12

This farmer has a sick patient. His diagnosis must be suffocation and not nitrogen deficiency. He must immediately get to the cause and not attempt to treat individual symptoms. If this were an animal, we would attempt to open its breathing passages. That is exactly what the farmer must do in this instance. He must recognize that recovery depends upon nutrient uptake which depends upon soil air. Therefore, his first action must be to introduce air through tillage as soon as possible. It is as if this soil had a plastic bag over its head and the corn will suffocate unless the seal is broken.

Nutrient treatments must be handled very carefully and not applied until the crop begins to show positive response from the effects of soil aeration. In other words, soil oxygen is the primary key to successfully salvaging this area. Soil oxygen supplies can range from adequate to totally deficient, or anywhere in between. As a result, yields can also range from high to low or anywhere in between. The management of soil air must be a top priority.

AEROBIC AND ANEROBIC SOILS

When a soil has an adequate level of oxygen we refer to it as being "aerobic." The term "aerobic" means free oxygen is present in some degree. A soil may be considered aerobic but still not have enough oxygen for maximum crop performance. An aerobic soil could be anywhere from as high as fifteen percent oxygen to low levels where crops will fail. For practical purposes we will refer to a soil as aerobic when the oxygen is sufficiently high to not be a limiting factor in crop production.

When a soil loses all of its free oxygen it then becomes what we call "anerobic." Under anerobic conditions all aerobic organisms cease functioning. Both the aerobic plant and the desirable aerobic soil micro-organisms die or become dormant. In addition to the aerobic soil micro-organisms there are also groups of anerobic organisms that actually thrive in the absence of soil oxygen. These same kinds of organisms thrive in septic tanks or liquid manure systems that are not aerated. Whether the soil is aerobic or anerobic, life goes on within the soil system. Anerobic soil conditions and the activities of anerobic organisms have a profound effect on the farmer's pocketbook and, for that reason, must be fully understood.

To understand the full effect of anerobic soil conditions, let us review what is happening in this condition, as illustrated in Figure 13.

When soil is actually under water, as in a pond, we notice that the plants thrive rather well for three to five days, depending on the temperature, and then begin to wilt and die. Rain water picks up a load of oxygen as it falls through the air and this oxygen-laden water

Fig. 13

fills the air spaces around the roots so the plant continues to function under a hydroponics condition until the oxygen becomes depleted.

A normal crop could be grown even if standing in water if oxygen could be maintained in the water surrounding the roots. In hydroponics the roots are placed in water and the plant thrives as long as air is bubbled into the water. But when the oxygen supply is turned off the plant will begin to wilt and quickly die.

Because water and nutrient uptake by the roots are oxygen-demanding processes we know that if the oxygen in the soil water is depleted the plant will wilt even though it is standing in water. Anerobic soil conditions guarantee crop failure.

In addition to crop failure anerobic soil conditions bring about another economic loss. Anerobic soil organisms are living creatures that must have oxygen to thrive. Because of the method in which they get their oxygen anerobic organisms cannot live in the presence of free oxygen.

The tetanus organism that causes "lock jaw" is a good example of how anerobic organisms function in relation to oxygen. Tetanus organisms thrive in puncture wounds that seal over and exclude oxygen. Tetanus organisms are found in the soil, but they remain dormant under aerobic conditions and cannot multiply in an open, aerated wound.

Anerobic soil organisms take their oxygen from oxygen-bearing compounds such as nitrates (NO_3) and sulphates (SO_4). The removal of oxygen from these compounds by anerobic organisms changes them to nitrous and sulphide gasses which escape from the soil into the atmosphere. Anerobic soil conditions can cause considerable losses of both sulfur and nitrogen. Without an understanding of what happens we will not be aware of losses because these gasses are invisible as they escape from the soil. Nitrogen losses through denitrification can be especially severe when soils become low in oxygen.

Anerobic soil organisms also take their oxygen from certain iron compounds. We refer to their end product as a form of "reduced iron." Reduced iron can cause several reactions because many plants tend to overfeed upon it. Excessive iron intake can darken and plug the nodes of corn, as shown in Figure 16. This plugging of the nutrient channels progressively slows the upward movement of nutrients into the plant.

Plugging of the nodes becomes critical during the ear-fill stage of growth. The plant becomes more prone to stress and the ear fill deteriorates, often showing curved, poorly-filled ends that reflect phosphorous and potassium deficiencies. High soil levels of these nutrients become meaningless if plugged nodes prevent their uptake and movement within the plant.

Progressive plugging of corn nodes causes the plant to age prematurely and die. The inner nodes will remain high in moisture and sugars inviting the invasion of soil fungi and stalk rot. The interior of a healthy, maturing cornstalk will not backlog water or sugars and will have the appearance of being filled with dry STYROFOAM*. Node plugging and stalk rot reflect a breakdown in the soil-plant relationship.

The excessive intake of reduced iron creates a unique problem when the forage is fed to cattle. Compaction or low oxygen may result in high levels of iron in forage. Forage grown under tight soil conditions can contain as much as 250 PPM or even more of iron. When this is fed to cattle the high iron will tie up phosphorous in the digestive tract, thus creating a phosphorous deficiency in the animal. The iron phosphates end up in the manure pit.

Low soil oxygen levels will also reduce the uptake and forage levels of other nutrients such as potassium and magnesium. Many a dairyman is perplexed to find high iron and lower levels of other elements on the second cutting of hay as compared to the first cutting. This can be caused by soil compaction and the resulting loss of oxygen which were imparted in the cutting, swathing, baling and fertilization trips while harvesting the first cutting of hay. Machinery compaction can be severe, especially on heavier soils if the farmer is forced to harvest when soil moisture levels are high.

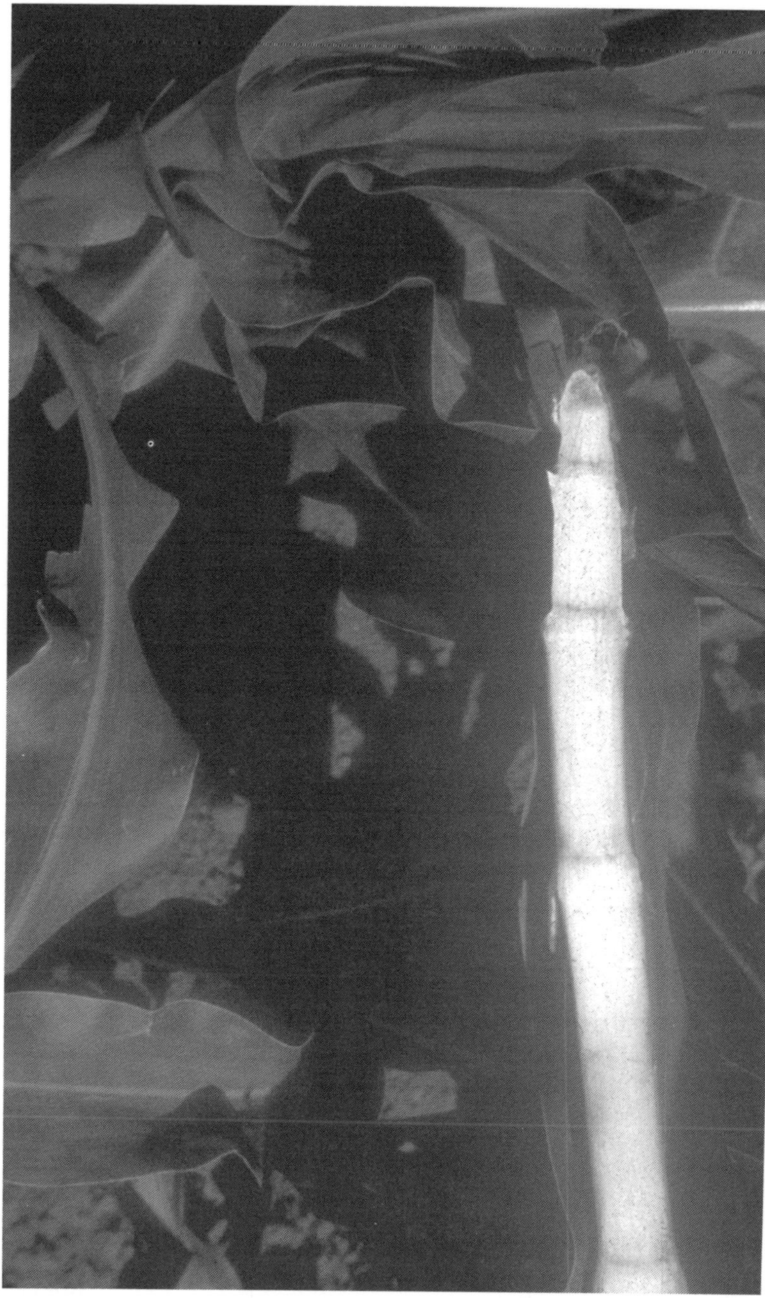

Fig. 16

Complete feed analysis and balancing of the feed rations for each cutting will become a top priority when the effects of fluctuating soil oxygen levels on the nutrient levels of forages is realized.

A lack of understanding and the improper management of soil air are the most limiting factors in production. It is impossible to estimate the total loss in production due to mismanagement of soil air . . . suffice to say, it must be staggering. In this study of **air, water, decay and fertility management** we must place first emphasis on air management, because without it, water, decay and fertility cannot be managed.

Oxygen is free. It also represents free fertility. Visualize oxygen as taking the nutrients by the hand and leading them into the root. There is no other way to get nutrients into the roots. Nutrient and water uptake is an oxygen-demanding process.

BASIC NUMBER TWO — MANAGING SOIL WATER

> In the management of soil water we must be concerned with the supply of usable water as well as its efficient use within the plant. The supply side of water management involves the utilization of the soil's total storage capacity. The efficient use of water involves producing a unit of production with a minimal amount of water. This becomes a function of adequate soil aeration and balanced soil fertility. Adequate soil oxygen and balanced nutrients give roots more suction or recovery capacity and literally turns them into powerful vacuum cleaners. Efficient water management is within the control of the farmer and is influenced to a great extent by his tillage methods.

Water is just as necessary to the well-being of plants as nutrients. In fact, we may consider water a nutrient and more. It plays several important roles in the life processes of a plant. As with an animal, much more water passes through the plant than it contains. It serves as a coolant by transpiring, or sweating, off of the leaf surfaces. It carries nutrients to the root surfaces and in this way increases the availability of the nutrients to them. Water is the universal solvent and essential to all life processes.

While water is necessary, it can also be detrimental. Excessive water can drown animals and suffocate plants, therefore it must be carefully managed. In many parts of the country, such as the "corn belt," we often suffer from drought even though more rain falls during the growing season than is actually needed. In general we can conclude that we are extremely wasteful with water. We permit too much to escape, creating problems of erosion and pollution.

Because water loss through runoff is a major concern we must improve our management of it. The efficient utilization of soil water is influenced by such factors as root size and depth, soil nutrient balance, recovery of water from lower depths back into the root zone, soil surface management, soil aeration, microbiological activity, and possibly even the position of the moon with its lifting effect on land and water.

Basically, water management includes the **capturing, storage, recovery** and **reduction of the total amount of water required to produce a pound of dry matter.** These are the four keys to management of water.

In the process of capturing, storage and recovery there are three important rules to follow. These are illustrated in Figure 17.

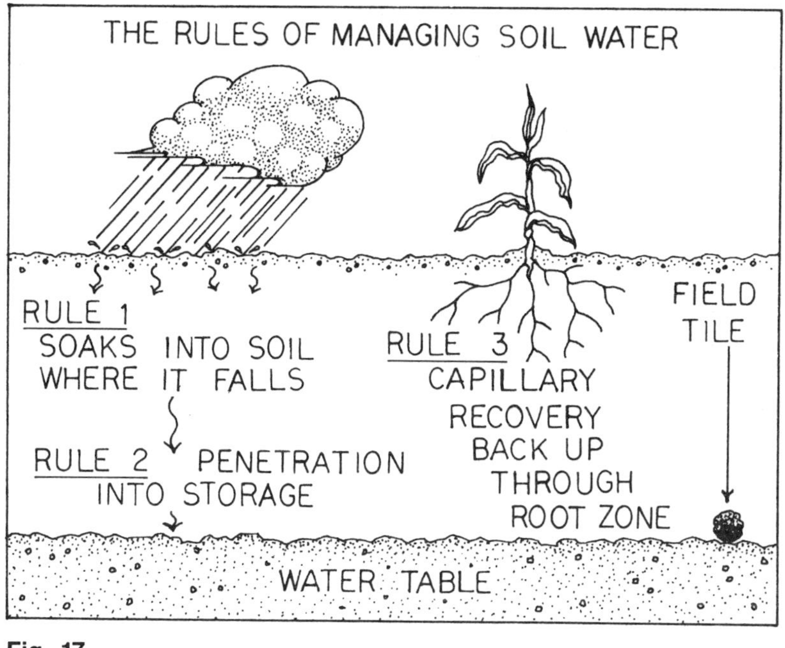

Fig. 17

Rule 1. We must make certain that rainfall or irrigation water soaks in where it falls. Runoff must be absolutely minimized.

Rule 2. The insoaking water must be able to penetrate the soil layers to reach the water table below. The height of the water table may need to be controlled by tile in some instances.

Rule 3. As we lose water through surface evaporation and leaf transpiration the lower water in storage must be able to rise by capillary action back up and into the root zone to replace the losses.

While water management principles are quite simple and involve only three phases, each phase must be thoroughly understood so that our soil management practices complement our water management principles. Often soil management practices hinder water management so severely that we suffer the effects of drought even though rainfall has been more than adequate.

SURFACE INSOAK

Phase One in water management concerns getting the water to penetrate where it falls instead of running off. Our concern is with surface conditions and how the soil surface is affected by the falling raindrops. A bare soil surface with no cover is the enemy of the water-insoak process. Falling droplets are fluid bombs moving soil in a manner similar to the exploding force of air. The difference lies in the ability of water to sort soil particles. Water can suspend soil particles long enough to permit a sorting either by particle size or weight. The larger particles settle out first and the finer particles settle out last and end up on the surface. This sorting, ending with the fine particles on top, results in the formation of an effective soil seal.

When these sorted particles dry out we notice a distinct crust that can vary from one-fourth to over one inch in thickness, depending on the soil type and the duration and intensity of rainfall. The sorting of soil particles, which leads to a sealing of the surface, guarantees runoff. Runoff is a leading cause of the loss of both water and topsoil.

The surface must be protected. This can consist of residue, living cover crops, crop canopy, or a combination of all three. Cover crops and crop canopies break the force of falling rain before it strikes the soil, easing the water on to the soil so there is no explosive force to separate soil particles. Crop residue, properly managed, serves as pipes or wicks to assist the insoak process and also to serve as a shield between the soil and falling rain.

WATER PENETRATION INTO SUBSOIL

Once water soaks through the surface it must be able to penetrate downward as far as necessary. The rate or speed of insoak will depend on the soil type and structure.

Soil texture is an important factor. Sand and coarse-textured soils hold less water per foot in depth, but absorb water much faster than fine-textured silt or clay soils. Subsoil structure also plays a role in the heavier soils. Some clay subsoils have a blocky, or cube-like structure. Others have a columnar structure consisting of vertical columns of clay. These soils take in water easily as it passes down the crevices or fracture lines.

Other factors, such as deep compaction from machinery traffic, plow or disc pans and natural clay barriers, can greatly influence insoak. Because these things are within our control they must be considered in our farming practices.

Most medium to heavy-textured soils that have remained in place for thousands of years have a natural clay barrier located just under the topsoil layer. These soils usually contain a combination of sand, silt and clay. As shown in Figure 18, clay particles are extremely small

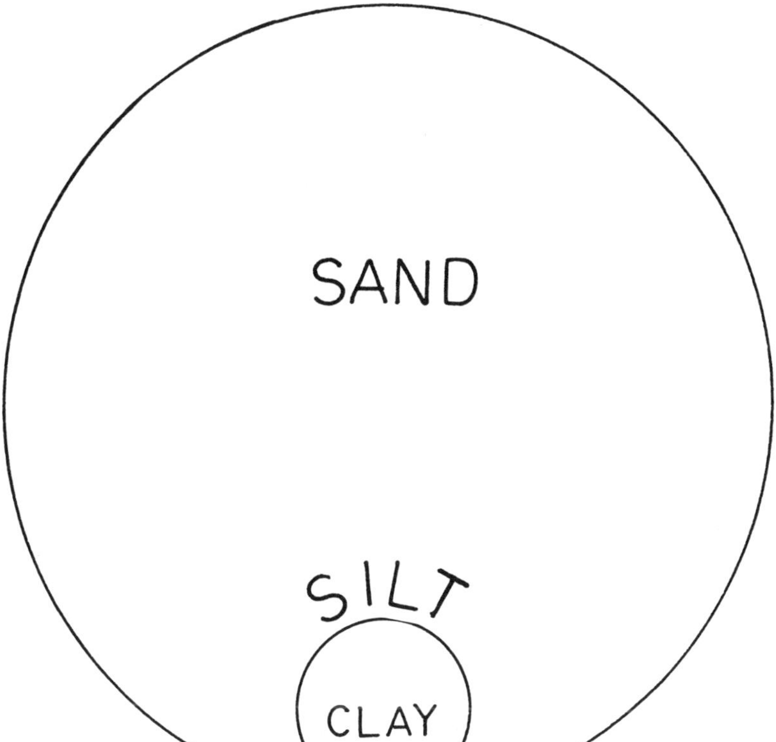

Fig. 18

in comparison to sand and silt. Clay particles are so small, in fact, that it requires the aid of a microscope to see them. Due to their tiny size clay particles sift downward over a long period of time forming a concentrated layer (barrier) below the topsoil. The clay layer's bottom depth depends on the age of the soil, percentage of sand, silt and clay and annual rainfall.

Figure 19 shows how the clay might appear in a middle-aged soil. Most of the midwest soils are middle-aged and have developed a "middle-aged spread" of clay. When these soils reach thirty to forty thousand years old, they will have reached "old age" where all of the clay from above will have settled into a solid clay pan. As soils age they become less and less productive.

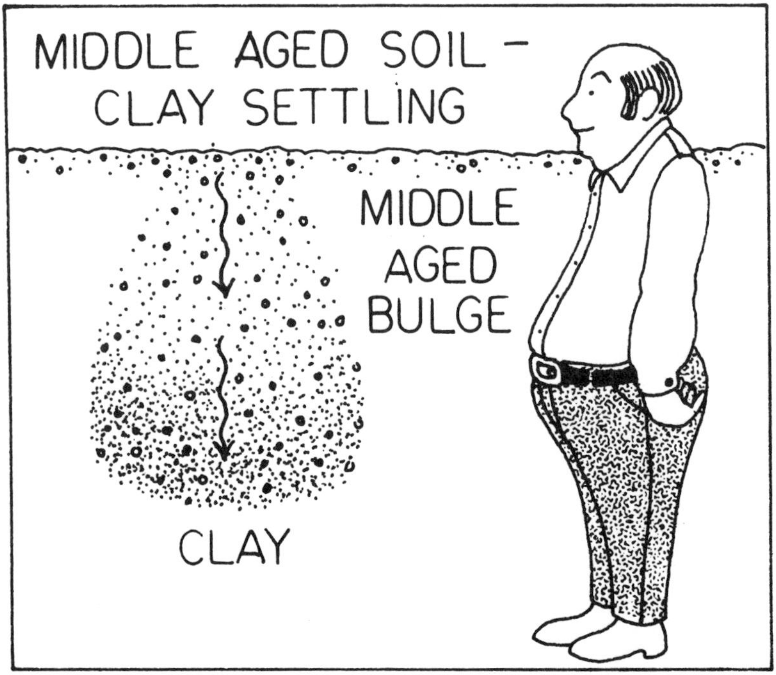

Fig. 19

Most middle-aged soils have a natural "clay zone." These soils can also develop plow pans and disc pans. Figure 20 illustrates clay distribution on a virgin soil and where a plow pan forms. The bottom of the moldboard plow will be within this high clay zone.

The bottom of a moldboard plow, as shown in Figure 21, that smears through moist or wet clay creates the same effect as running a trowel over fresh concrete. It smears the colloidal clay surfaces and imparts a compaction zone from one to two inches in depth. This is called the "plow pan." The plow pan, with its smeared surface, forms a distinct separation between the plowed zone and the lower soil horizons. A plow pan is nearly void of air.

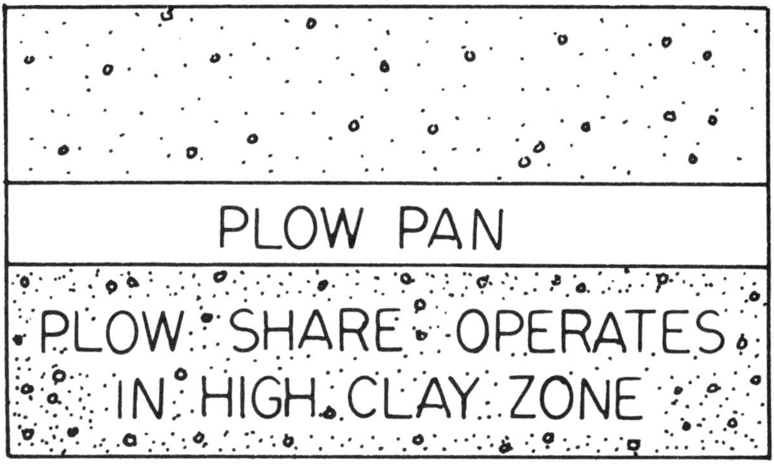

Fig. 20

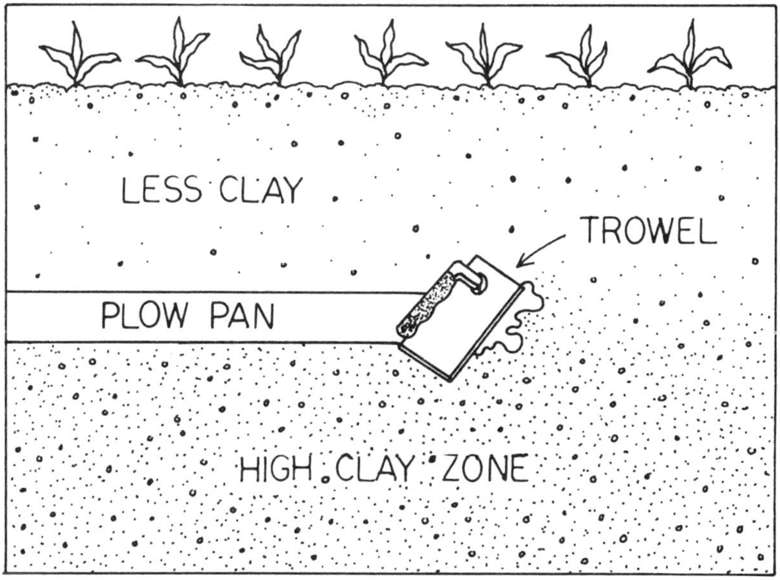

Fig. 21

These pans are very significant in the management of water. First, they resist and greatly impede the downward movement of insoaking water, often causing the plow layer to become totally saturated and void of air because the water cannot move down. The insoaking water must push air with it if it is to penetrate. If the air cannot move through the pans an air lock will develop which, in effect, will hold the water up. The overall effect of these pans is to prevent water from penetrating into the vast storage area below and, as a result, surface runoff will be increased.

Because of a lack of oxygen these dense pans prevent the downward movement of roots. In effect, this limits the depth at which we farm and restricts the feeding area for both water and soil nutrients. We may view the topsoil and subsoil as a two-layered cake. What a waste of potential it is to farm only the top layer!

Soils hold approximately one to three inches of water per foot of depth. Obviously, more water will be available to the plant as roots penetrate deeper into the soil.

Figure 22 shows an interesting situation that developed on a wheat farm in Ontario Canada. This field, as with several others around it, had prospects of a very good wheat yield. However, to the amazement of the farmer, the wheat suddenly died during the early milk stage, while the fields near it looked healthy. While investigating the problem we noticed that this field was laced with deep and wide cracks while the surrounding fields of the same soil type displayed no cracking. The cracking clearly indicated that this field had dried out suddenly. A sudden dryout suggested that the water losses above were not being replenished from below. That is what had happened.

Fig. 22

The block of soil shown in Figure 22 was easily pried out with a spade. The block extended to the depth of the plow layer which in this case was only five to six inches. The base of this block was covered with an organic blanket. The farmer remembered that he had plowed under a very heavy crop of red clover ahead of fall seeding of wheat. This organic sandwich separated the plow layer from the lower layer. Because the continuity was broken and the capillary rise of water was stopped by a plow pan and an organic layer, the shallow plow zone had dried out. Due to the plow pan and organic sandwich this farmer was farming only the top layer and in this case it just was not enough.

There was adequate moisture in the area: all of the fields around produced well; it was simply a situation of poor management of water which had been brought about by poor and improper tillage procedures. A costly lesson was learned the hard way.

One of the goals in soil management is to maintain continuity between the topsoil and the lower horizons or layers. This requires the elimination of organic layering, excessive machinery compaction and natural or tillage induced pans. With the wide use of the moldboard plow and intensive row cropping much of our most productive soils suffer from organic layering, plow and clay zones. With the technology available today it is difficult to understand why we would consistently place such severe limitations on potential production. It is almost as if we were trying to prevent surplus production!

Lowered production is serious, but takes a back seat to the much more serious and long-lasting problem of soil losses because of surface runoff. Problems that prevent insoak must be identified and eliminated. Properly incorporated residue instead of layered material can be accomplished by adjustment of tillage practices.

While clay or plow pans are out of sight, we can usually see their effects. Standing water, as shown in Figure 23, indicates poor under-

Fig. 23

drainage which could have resulted from a faulty tile, a naturally high water table, or pans that have prevented the water from moving downward. In this instance both a plow pan and a natural clay zone are keeping the water from penetrating into the subsoil. As a result the tile in this field is not able to function as intended.

Conditions such as these can cause a multitude of problems in certain years. A wet fall can cause harvesting problems, damage the soil with ruts as deep as the pans and often prevent any fall tillage. For this field a normal spring then turns into one too wet, resulting in late planting, working the field when it is too wet, causing clods, more problems with insects and weeds, losses of nitrogen and sulfur through gasification, extra expense in tillage, finally to be capped off with lowered yields.

Given the detrimental effects of clay or tillage pans, how can they be managed? We approach the matter in terms of seven major questions. (1) Where are the pans located? (2) How can they best be removed? (3) How deep must we till to remove them? (4) What tine spacing is best? (5) How often will sub-soiling be needed? and (6) When is the best time of year to sub-soil?

A plow or clay pan can be located by using an ordinary soil probe. Plow pans begin at plow depth and can be one to two inches thick although natural high clay zones may be as much as eighteen inches in depth depending on the erosion of topsoil. Probing should be done when soils are moist or near their full field water-holding capacity. Wheel tracks must be avoided in selection of the probe sites. The pans are located by noting pressure changes as they are intercepted, and also by thumbing the soil core to feel the differences in intensity.

Once located the pans are best removed with the practice often called "subsoiling" or "deep chiseling." The operation uses a narrow shank that will penetrate sufficiently but leave the lower clay in place instead of bringing it to the surface. There are enough potential problems in the seedbed zone without the addition of more clay. The three or four-inch twisted blades should not be used to remove pans because they roll the lower clay back up to the surface. A rigid-framed V-plow, as illustrated in Figure 24, may be ideal. Ideally a subsoiler should be selected that has the ability to handle residue and shanks that can trip and reset. Any type of shank that will penetrate and smear a channel through the moist clay pan while leaving it in place will be satisfactory.

Usually there is little need or benefit to deep till more than two to three inches below the clay zone. Power costs are too high to go any deeper than is absolutely necessary. However, the subsoil tine must penetrate the problem zone to solve the problem.

Fig. 24

Spacing of the tines is another consideration and there may be no single spacing to ideally fit every situation. Experience indicates that a thirty-inch tine spacing does a good job. Closer spacing demands a big increase in power and does not show any conspicuously better results. A balance between the power costs and solving of the problem must be struck.

In nearly all situations deep chiseling to remove problem zones should be done in the fall. Medium to fine-textured soils that develop pans are the soil types most easily damaged if worked too wet in the spring. Wheel track compaction must be minimized in the spring. Because of the tractor size required for the subsoiling operation and the narrow width of the deep working tool the field could become literally covered with deep, packed tire tracks. One of the major benefits of deep tillage is to improve water management and reduce tillage damage in the spring.

The question, how often must I sub-soil, is an important one and impossible to answer precisely. Removing the pans is only one phase of a complete tillage system. A plow imparts a pan every year and can nullify the sub-soiling operation. For this reason alone the plow should be eliminated and not considered a tillage system. If primary tillage is done properly, as outlined in the unit on tillage, a sub-soiling will last almost indefinitely, depending upon the soil type and seasonal conditions. The best guide is often the observations of the farmer concerning insoak, how well field tile is functioning and spring working conditions. Some farmers sub-soil their end rows every year even though the remainder of the field does not need it.

Subsoil slots have a tendency to stay open for long periods for several reasons. The roots of plants will penetrate these slots and eventually build a solid wall of organic wicks or tubes. This helps to maintain the passageway. Also, expansion and contraction from the wetting, drying, freezing and thawing seems to help keep a slot line open. The same expansion and contraction has little or no effect on removing or breaking up a solid band of clay or plow pan. A solid band will not fracture because it is under pressure on all sides.

Figure 25 illustrates the method and results of proper subsoiling. The slots every thirty inches permit the penetration of both roots and water. They permit water to move upward from storage into the root zone to replace the losses from evaporation and transpiration. They build continuity between the topsoil and subsoil and permit the farming of the whole "cake" rather than just one layer of it.

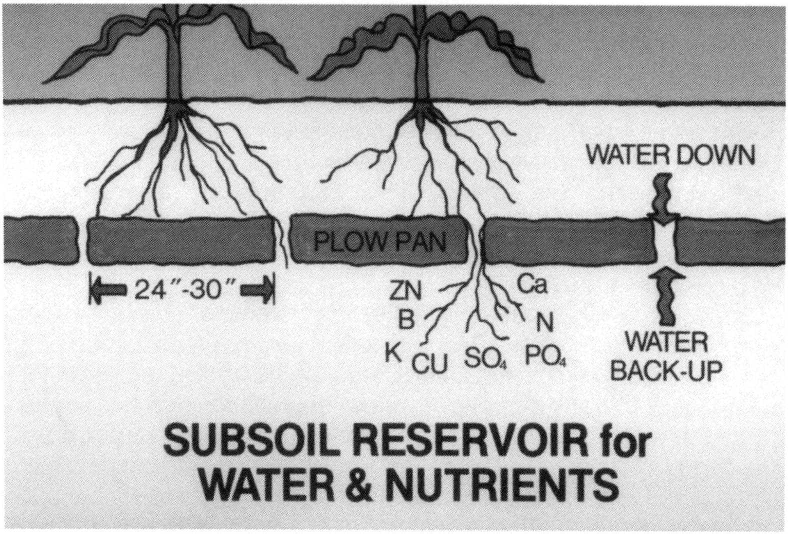

Fig. 25

Figure 26 shows an interesting subsoiling experience on a droughty soil. Even sandy soils can have clay or silt pans. This sand had a tight, two-inch thick pan about sixteen to eighteen inches below the surface. The farmer subsoiled a couple of strips to see if there would be a difference and that particular year turned out to be very dry. Notice how much taller the corn is over the subsoiled strip. A trench showed the corn roots growing through the deep tilled slots and continuing downward another twelve to eighteen inches. This is what is meant by continuity and farming all of the layers of soil. It is what subsoiling is all about.

A few points about subsoiling must be clarified. First, do not assume that your whole farm needs it. Some soil types do not have

Fig. 26

a problem with clay and plow pans. If one-third of a farm is trouble-free there is no need to waste power and time by chiseling that part of the farm. The purpose of probing is to determine if there is a problem and its extent. **Your county SCS personnel can be your best resource people to help you in determining the extent of your problem.**

Second, do not assume you must shatter the subsoil. A clay subsoil can be shattered if the clay is dry. It would be rare to find this lower depth completely dried out. If you waited for this it could be several years before these lower zones could be shattered with subsoiling. If you do sub-soil when it is dry it will require more horsepower and chisel points will wear out very quickly. It is also questionable whether sub-soiling under dry conditions will endure. A clay zone shattered when it is very dry will expand and close when it takes on moisture. Cracks that form in clay soil during extended droughts may be two or more inches in width and extend two to three feet into the soil. While it would appear that these cracks provide free sub-soiling, in reality when the dry clay takes on water it expands the clay and closes these cracks as tightly as they were before they dried out.

When sub-soiling in moist clay we smear the sides of the slot and in effect put a vertical plow pan on each side of the slot. As the tine moves through the moist clay zone it will wedge the soil apart and the slot will remain open. Topsoil will silt into these slots and within two or three years they will fill with a wall of roots growing into the subsoil. Roots will break through and penetrate the lower soil stratus. Water will move downward into storage and by capillary action move back up as the root zone loses its water. In effect sub-soiling is a means of "weatherproofing" crops. We are farming the entire soil column and not just a zone above the plow pan.

The sole purpose of sub-soiling is to break through plow pans or natural clay zones. Primary and secondary tillage involves the working of the soil above these problem areas, while sub-soiling works the soil to sufficient depths to remove problem zones. Distinguishing the difference helps when we think in terms of a total tillage program.

WATER RECOVERY FROM SUBSOIL TO ROOT ZONE

The study of soil water management would not be complete without a discussion of water storage and its movement within the soil. Figure 27 makes a comparison between an oil lamp and the soil. The oil from the storage reservoir is drawn up the wick to replace the losses from burning. The losses from burning can be compared to the losses of soil water which result from surface evaporation and leaf transpiration. When the lamp runs out of oil the fire goes out, and so it is also with the plant and water.

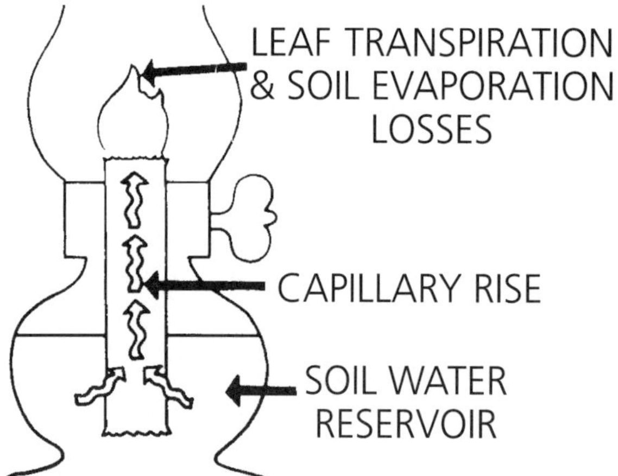

Fig. 27

In Figure 28 we show how water is held and moves within the soil. Soil holds water in a capillary film around each soil particle. A soil is said to be at "full field capacity" when each soil particle holds a full thickness of capillary film. At this point the soil feels moist, but not wet. It will not form a mud ball. At "full field capacity" it still has

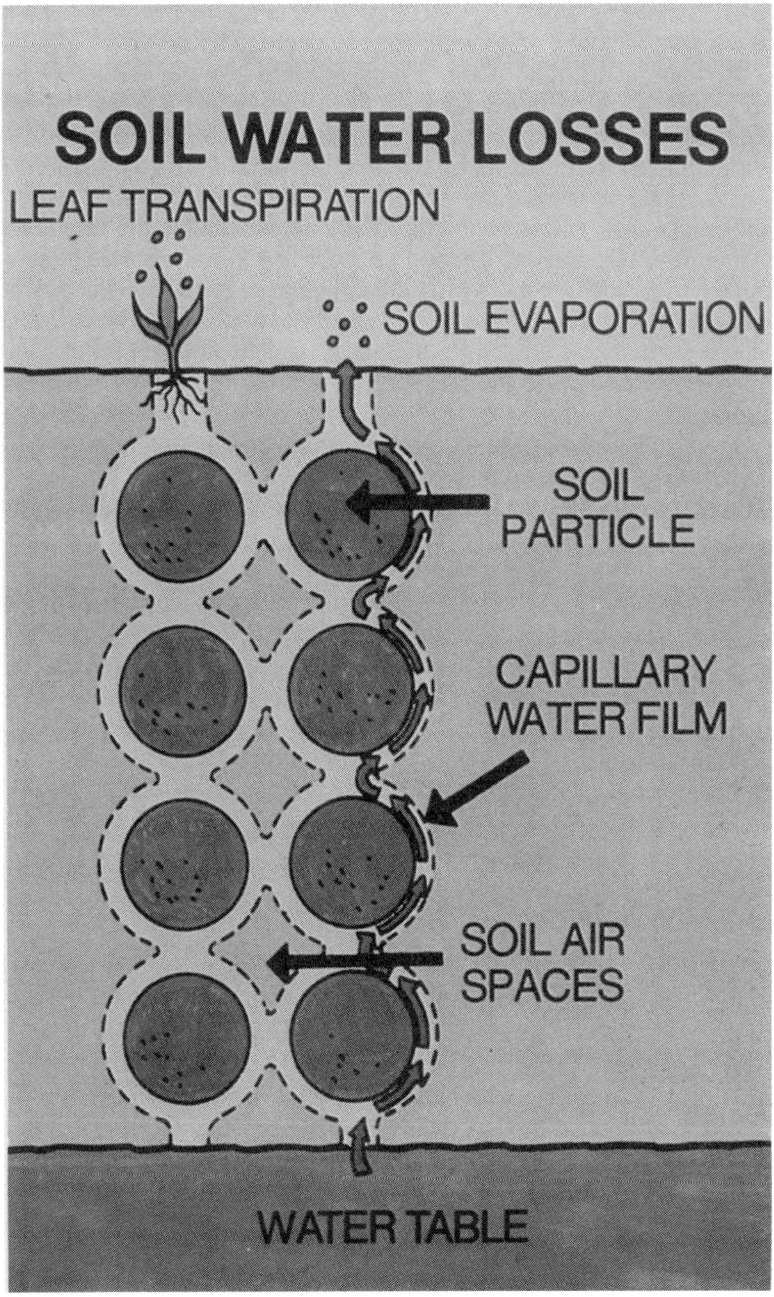
SOIL WATER LOSSES
LEAF TRANSPIRATION
SOIL EVAPORATION
SOIL
PARTICLE
CAPILLARY
WATER FILM
SOIL AIR
SPACES
WATER TABLE

Fig. 28

a full complement of soil air. If more water is added the capillary film is not able to hold it and the air spaces begin to flood with free water.

An interesting thing about a capillary film of water is that it tries to maintain equal film thickness from the water table to the soil surface. This is due to capillary attraction or surface tension. As the water is used from the upper soil film, the water from below will rush or be pulled up the column to replace the loss and to replenish the film thickness. This is an important principle because the losses will continually be replaced as long as there is a supply of water below. As the reservoir of a lamp must be maintained or the fire will go out, so also must the soil's reservoir of water be maintained. This places much importance on soil management practices that permit water not only to penetrate the surface, but also to move downward into storage.

Figure 29 stresses the extreme importance of water management by showing a soil particle surrounded by a full thickness of capillary film and penetrated by an absorbing root hair. The thickness of the

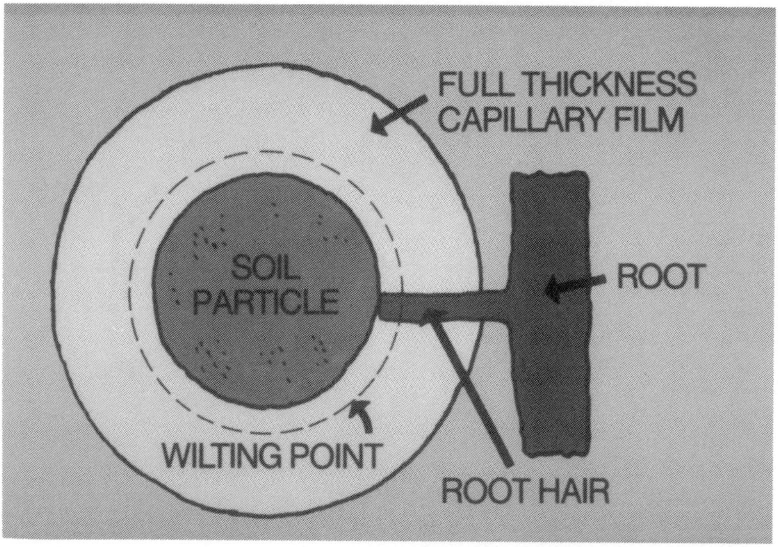

Fig. 29

capillary film determines how loosely it is held and how easily it can be extracted by the root. As the capillary film shrinks it requires more and more atmospheres of pull by the root to remove it. When the film shrinks to the point where the transpiration losses from the leaves exceed the ability of the root to remove the soil film, the plant reaches the wilting point. At night plants come out of the wilt because the losses from transpiration are less than during the heat of the day.

The permanent wilting point is reached when the plant cannot recover during the dark hours.

Also, remember that the wilting point is influenced by two factors. **Soil nutrient balance and soil oxygen determine the suction strength of the root.** As we have said before, soil aeration provides the oxygen which is necessary for the uptake of nutrients including water. For these reasons soil aeration and nutrient balance are a definite part of water management.

Good soil nutrient balance with adequate soil air to assure recovery results in a reduction of the quantity of water needed to produce a pound of dry matter. Water management must include capturing, storing, recovery and the efficient utilization of water within the plant.

BASIC NUMBER THREE — MANAGING THE DECAY PROCESSES WITHIN THE SOIL

> We would live in a very peculiar world if nothing decayed. Our planet would soon die. Decay is nature's way of securing life through recycling. Life and death are interdependent and one cannot exist without the other. This is especially true within the soil-plant system. The living soil and the plant need each other and they are both needed by humanity. In our quest to advance agriculture we must complement nature's system and not detract from it.

The purpose of this chapter is to gain an understanding of the importance of soil organisms and how to manage them for maximum benefit of the soil, plants and ultimately, ourselves.

Research in this area has lagged far behind other technological advances in agriculture. It is somewhat difficult to understand why we have spent so much effort and research in the study of the physical and chemical makeup of soils and so little effort in soil biology, which may well prove to be a most important aspect in relation to the welfare of both the soil and plant.

We often refer to "the living soil" and this is quite correct in view of the multitude of living creatures that dwell and function within the soil. Basically, the soil contains two classes of consumer organisms which are called "micro" and "macro." Micro-organisms are microscopic in size, while the macro-organisms are large by comparison. Earthworms, insects and larvae are good examples of macro-organisms. Both groups are consumers and participate in the decomposition or recycling of organic materials within the soil system.

Life on earth is a delicate balance between producers and consumers. When consumer demands for food exceed the supply, the natural law of balance simply demands adjustment of consumption. This adjustment proceeds through the simple process of starvation or malnutrition and disease, which in time will certainly adjust the demand for production.

The producing plant is also controlled by balance. A balance of nutrients is as necessary for the plant as it is for the consumer. If this balance becomes upset the plant must adjust its needs or else give way to a less demanding species by succumbing to malnutrition, disease and predation. Plants are in part consumers, because they must ingest nutrients before they can produce. If their nutrient or consumer needs exceed the supply they become deficient producers which has a great impact on all consumers whose nutrient needs must be satisfied either directly or indirectly from the plant source. The statement, "poor soils produce weak animals," is certainly a fact of life.

The soil-plant relationship is a fascinating story of adaptation and survival. It is very difficult to define the term, "soil." An early writer simply defined it as a place for plants to anchor their roots. As we learn more about the soil and plant relationship we must continually revise our definition. We realize that the soil teems with a great variety of living creatures who are consumers and we also know that these consumers are working in harmony with the living plant.

For plant life to develop and endure nature had to assure a continuous source of nutrients. To meet this need the lower forms of microflora had to evolve and develop in tandem with the plant kingdom. To this end it was necessary for nature to form a symbiotic dependence between the food-producing plant and the consuming organisms of decay. From this relationship the soil-plant system became firmly established. The consuming soil flora multiplied in species with specialized purpose and function that further developed into an extremely orderly intrafunctioning ecology, seemingly intent on providing for the total needs of the food-producing plant.

A greater intelligence seemingly relayed the message to these various organisms to put their house in order because their existence would depend totally on the welfare of the plant. A recycling of plant remains was absolutely essential to rid the soil of choking residue buildup and to provide an endless supply of balanced nutrients for both the producing plant and soil organisms.

A climax plant species dominated an area because it found both the soil and climate to its liking. Soil chemical balance varies for different plant species. The recycling of a dead plant returns back to the soil the ideally-balanced fertility package for that particular species. Recycling by soil organisms is the ultimate in both conservation and fertility balance.

Recycling of nutrients is the key to continued life. One wonders about our own body nutrients and how many times they have been recycled through other living forms before they became part of a person.

Since the plant is the producer it appears that the decomposition of all organic matter is for the purpose of providing for the welfare of the plant. Nature is telling us loud and clear that what comes from the soil to support life must return to the soil if life is to continue.

Because of nature's priorities the soil-plant relationship developed into a "closed system" that could endure all by itself without the higher forms of life above ground. It is interesting to observe that the consumers within the soil are not life-threatening to the plant as are the consumers above ground. Soil organisms consume only the dead remains of plants and in the process they support and defend the plant.

SOIL, THE STOMACH OF THE PLANT

Digestion is designed to break down complex substances into simpler components that can be absorbed and utilized by an organism. We may mistakenly assume that digestion applies only to organisms that have a self-contained digestive tract. Digestion simply makes nutrients available for absorption and need not be self-contained. Many microbes send their enzymes outside their bodies to break down substances into simpler forms that can then be absorbed back into their bodies. This is an excellent design that permits them to remain microscopic. Remaining small has advantages: it permits them to permeate into the tiniest places with minimal nutrient needs. Internal digestion demands size and a lot of energy to function.

Enzyme digestion by microbes also plays a role in plant nutrition. Enzymes are specific in their function and will break down a given substance without being destroyed in the process. This provides plant nutrients in two ways. First, the plant can use the enzyme-released nutrients and, second, it can absorb the enzyme as such. In other words, plants can absorb enzymes from this process which eliminates the energy-demanding process of building them internally.

It is important to know that plants can take in elemental as well as complex molecules. It is a serious mistake, however, to believe that a plant can thrive on individual elements alone.

It used to be quite simple to list the elements needed by plants as carbon, hydrogen, oxygen, phosphorous, potassium, iodine, nitrogen, sulfur, calcium, iron and magnesium. To help remember them we listed them in a given order (C-HOPKINS Ca Fe Mg). This conveniently spelled out the words, "See Hopkins Cafe — mighty good." We were also taught that a few trace elements should be added to this list.

If plants needed only the pure elemental nutrients there would be no need for decomposition of residue organisms. Nature could periodically send lightening bolts to earth to burn the accumulation of residue to release the pure elements for plant use. The fact is, if a plant is to produce its full genetically-designed nutrient package and remain healthy and free of disease and insects, it requires more than raw elements to satisfy its needs.

A plant does not have an internal digestive tract and must depend on external digestion and absorption through its root system. Therefore, the soil microbes substitute for the stomach of the plant and externally digest complex substances into simpler forms that the plant can absorb.

Figure 30 is designed to illustrate the differences and also the similarities between external and internal digestion.

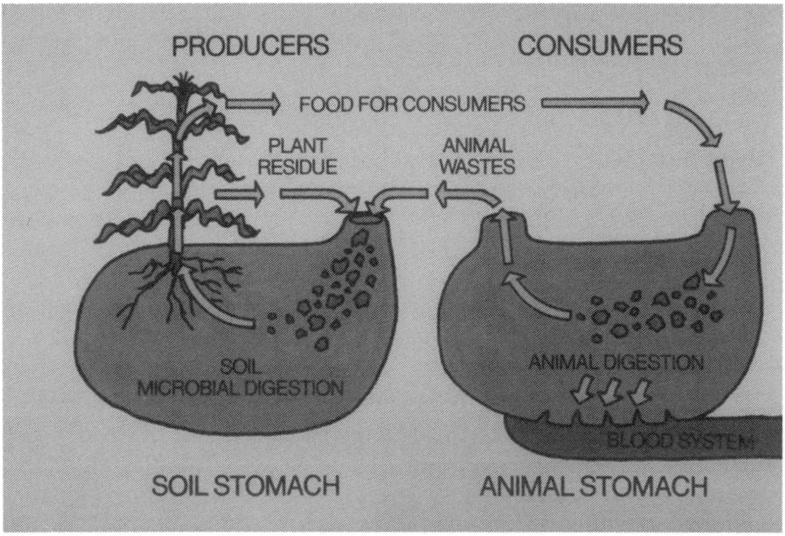

Fig. 30

An animal's digestive tract is designed to accept and break down complex substances into simpler forms that can be absorbed directly into the blood stream. In the soil the soil organisms perform this function so that the simpler substances of this digestive process can be absorbed through the root hairs and into the "blood stream" of the plant.

From this soil digestion comes forth much more than simple elements. Decay releases enzymes, hormones, antibodies, chelated elements and what we call "unknown factors" that cannot be precisely identified. Given what is known about the co-evolution of soil and plants, we must assume that the total decay process and end products must certainly relate in a positive manner to the living plant.

This does not infer that we cannot grow plants on pure elements alone. This is done with hydroponics. We can insert the pure elements in water, oxygenate the water so the roots can absorb the nutrients, and grow a plant. However, there is a big difference — the true test of the plant's quality can only be determined by its effect on the growth, development and health of the consumer.

If the effects of feeding both hydroponic plants and those grown naturally on a fertile, living soil were compared, we would notice a big difference in animal performance in favor of the latter. If we compare flavor and the plant's ability to resist disease and insects we should note an advantage in favor of the naturally-grown plant. Man's technology cannot favorably change nature's design.

It seems simple to analyze the chemistry of an organic substance. A good chemist can accurately determine all of the elements in an apple. But let us artificially encapsulate those different elements into a pill. Would we expect the pill to "perform" in an animal the same as an apple? It would not, because there are factors and combinations within the apple that cannot be duplicated by man.

BENEFITS OF DECAY

The decomposition of organic materials provides many by-products and beneficial aspects that relate directly to the plant. Humus or organic ash is the end product from the biological break-down of carbonaceous residue. Humus acts as a cushion or shock absorber surrounding the soil colloids. This cushion prevents soil particles from touching and as a result reduces soil compaction, or bulk density.

Humus increases the soil's pantry size or fertility-holding capacity. It has far greater cation exchange capacity than the highest CEC clays. Increasing humus is the only practical way to increase a soil's pantry size.

Depletion of humus is a universal problem in modern agriculture. Our richest prairie soils may have lost as much as seventy-five percent of their original humus levels. As a result our soils suffer increasing erosion, lowered water and fertility-holding capacities, increased bulk density and reduced aeration and a reduction in yields. Our goal in soil management is to stabilize humus and begin the slow process of rebuilding.

Lowered humus on medium to heavier-textured soils makes it more difficult to manage air, water and decay. This in turn compounds fertility management. However, through correct tillage and residue management we can reap the benefits of decay while slowly building humus.

The decay process releases several mild organic acids that etch out elements from soil minerals which can contribute substantially to fertility maintenance. The decay process organically chelates nutrients, which maintains their availability and improves their uptake and use by the plant. Hormones, enzymes, antibiotics and no doubt a host of unknown factors constantly being discovered by research are also contributions of the decay process.

Finally, decay organisms condition the soil in a most unusual method. The exudates of their bodies cause a clumping, or granulation, of soil particles resulting in improved soil structure, tilth and aeration. By this process soil organisms construct their own soil ventilation system that assures their air supply. The increased air supply is very beneficial to the roots of plants as well. This granulation process partly overcomes the effects of low humus. Humus increases the stability of good soil structure and makes tilth easier to maintain.

We must be optimistic about our capacity to reap the benefits of decay rather than lament depleted humus supplies as an irreparable loss. If we properly manage decay we will not only recognize immediate benefits, but will also begin the long-term process of rebuilding humus.

Two principles concerning soil microbiological activity should be thoroughly understood by everyone concerned with the science of growing plants. These two principles relate to "nutrient competition" and "zones of biotic activity" within the soil. Understanding these will clearly define some existing problems and give a clear insight to their solutions.

THE FENCE POST PRINCIPLE

Where does a fence post rot off? Nearly everyone will answer, "near the soil surface." This is correct and understanding the fence post principle will help us to know where crop residue must be positioned if it is to decompose properly.

Figure 31 shows that an untreated fence post will decompose starting at the surface and extending a few inches below the surface. Below this point there is little decay on the post. Figure 32 also illustrates that under natural conditions dead plants are returned to the surface of the soil to decompose. Nature intended residue to decom-

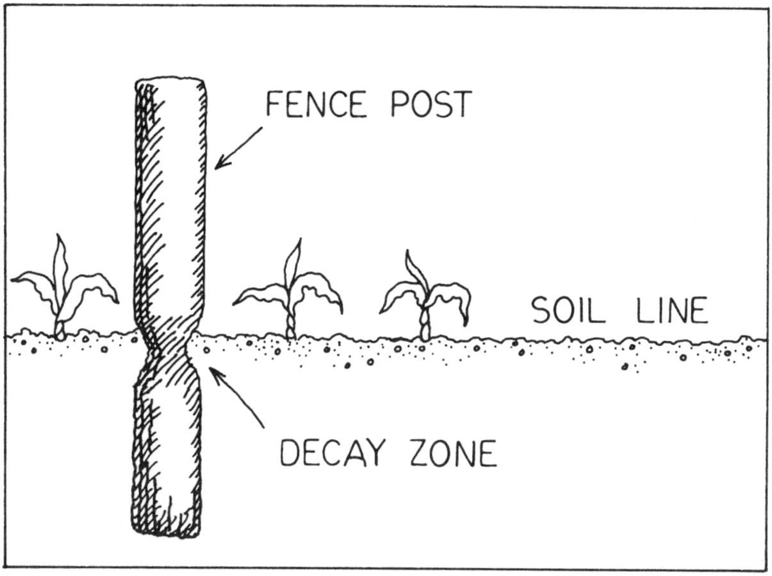

Fig. 31

pose on or near the surface. The significance of this will be explained in the second principle relating to decay.

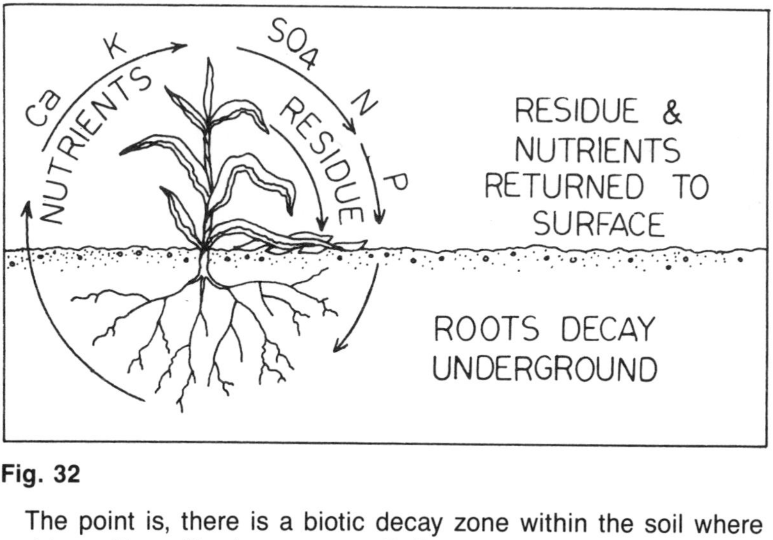

Fig. 32

The point is, there is a biotic decay zone within the soil where residue will readily decompose. Soil organisms require adequate oxygen and soil temperature. These two conditions deteriorate as we move deeper into the soil. Surface decomposition is part of nature's plan. When we plow or dig up residue that was buried a year ago we must recognize that we have run afoul of the natural system. We simply buried the residue deeper than where a fence post could decompose.

Figures 33, 34 and 35 illustrate what happens when residue is buried too deep. The heavy sudax crop in Figure 33 was plowed under during midsummer. Figure 34 shows the sandwiched material

Fig. 33

Fig. 34

still not decomposed after ten months. Figure 35 shows wheat stubble returned to the surface undecayed after eight months underground. This deeply-placed residue will partially undergo a form of decay that can release alcohols, aldehydes and formaldehyde compounds that preserve them further. These compounds can also be toxic to plants. Sandwiched, undecayed residue presents other problems. It interferes with the capillary rise of water, as well as water penetration.

The toxic release of chemicals can produce an unfavorable environment for roots. In addition, we lose the many benefits derived from the decay processes and tie up the nutrients within the residue that could otherwise be released and utilized by the following crop.

Fig. 35

COMPETITION FOR SOIL NUTRIENTS

The second principle to note is that decay organisms within the soil can compete with the growing crop. Figure 36 illustrates this.

As the soil warms soil life begins to increase. If there is food available these organisms continue to build in numbers until their food source is consumed. As with all living creatures, they begin to decline when their food source diminishes. The living cannibalize the dead on their way down.

As the microbiological numbers build, the chart shows that certain soil nutrients decline. This is the direct result of soil nutrients being tied up within the microbial bodies. We are mostly concerned with nitrogen, sulfur, phosphorous and some of the trace elements.

Nutrients that are in either the residue or the microbial bodies are unavailable to the growing plant.

The chart shows also that as the microbiological numbers decline the soil nutrient level rises to higher levels. This increase in fertility levels results from the release of all of the nutrients that were in the crop residue and microbial bodies.

Nutrients within crop residue can be released either through decomposition or by burning. Burning should be avoided because it prevents the building of humus and eliminates the many benefits derived from the decay process.

Decay organisms are consumers and Figure 36 drives home the point that we must manage the decay cycle so that its peak activity does not coincide with peak crop production periods. We do not want the ear of corn to be filling when soil nutrients are at their lowest point.

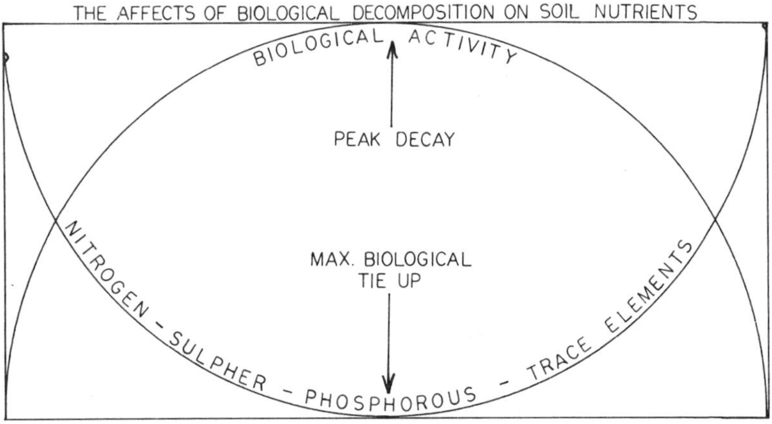

Fig. 36

If soil organisms and the crop are permitted to "sit at the same dinner table," the crop will always lose. Figure 37 shows earthworms, white actinomycetes, plus a multitude of invisible organisms very busily dining on residue.

The little corn plant in Figure 37 can survive this voracious competition because its needs at this stage in growth are not large. However, if this activity were to take place during pollination and ear fill the corn yield would suffer. It would probably be shocking to us if we really knew how much yield is lost because of competition from decay.

An interesting observation is that nature did not allow decay to compete with plants. Figure 32 shows that, under natural conditions, plants died and decomposed on the surface. This served several

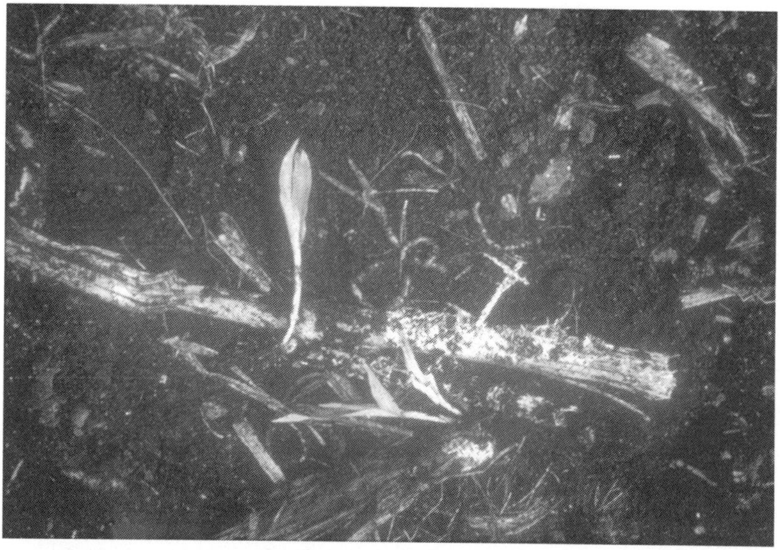

Fig. 37

purposes: it recycled lower nutrients back up to the surface where they were released through decay to slowly seep into the root zone. Of importance, however, is the fact that surface decay did not compete with the growing plant. Nature did not permit competition.

To prepare a seedbed for agricultural crops we must incorporate residue through the process of tillage. We incorporate residue throughout the root zones which automatically sets the stage for competition when it decays. Our goal must be to complete the decay cycle during the earlier stages of plant growth when the crop needs are less.

A few things can be done to speed the process of decay. First, we must properly incorporate the residue within the biotic zone of decay "where the fence post decomposes."

Second, we must recognize that carbonaceous residue such as cornstalks, grain straw and sunflower and cotton stalks are difficult to decay. This type of residue is high in carbon and low in nitrogen. A cornstalk may contain sixty parts of carbon to only one part of nitrogen. Soil organisms, in similarity with other animals, use carbon for energy. There is enough carbon in an acre of cornstalks to build a mountain of organisms, but there is not enough nitrogen to build their protein bodies.

The addition of nitrogen, along with a little phosphorous, can greatly enhance the decay process when residue is properly incorporated. However, it does little good to apply these to residue that is placed below "where the fence post decomposes." This is a waste of resources because the residue will be plowed back up the next year still not decomposed. Recommendations for the correct use of nitrogen and phosphorous for helping decay will be outlined in the chapters covering tillage.

Decomposition of crop residue is accomplished by a host of different soil organisms from bacteria to fungi. The actual number of different types of organisms functioning within the soil mass is greatly influenced by the kind of residue being decomposed. It requires a full spectrum of organisms, including bacteria, virus, actinimycetes, protozoa, fungi, algae, yeast, and other micro and macro-organisms such as earthworms, all interacting for the maximum welfare of the soil-plant system.

Interaction between the many kinds of soil organisms benefits the plant in many ways. Certain fungi, such as the endegones, attach to the roots of some plants, such as corn, and function as root hairs in bringing nutrients into the plant. Certain nemataphagus fungi and soil protozoans may actually consume some detrimental nematodes. The production of antibiotics and the myriad interactions between the many soil organisms can provide a wall of defense for the living plant. With a balance of soil organism activity, coupled with a healthy plant's own defenses, the plant has little to fear from disease and insects.

Carbonaceous materials, such as corn or small grain residue, will not necessarily proliferate the desired spectrum of soil organisms. "Straw stack bottom" type of decay, by itself, will not satisfy the needs of a healthy soil-plant relationship. While animal manure is beneficial, carbonaceous residue must decompose along with a **green material** such as rye, a legume, or even weeds, for maximum effect and benefit to the soil-plant system. Broad spectrum decay is a soil-cleansing process that biodegrades organic remains and many man-applied toxins as well. Considering the host of toxins made available through technology, but foreign to nature, we need as many soil cleansing processes as possible.

A growing crop stimulates soil life before and after death. While living, the roots attract and nourish many symbiotic soil organisms along their surfaces. After death the plant supplies an available source of nutrients to further stimulate the proliferation of soil life.

The quantity of energy captured by growing plants regulates the quantity and quality of all life. The non-plant biomass underground probably greatly outweighs the non-plant biomass above ground. The varieties of underground biomass influence the quality of plant life, which in turn determines the health of all other forms of life that consume either directly or indirectly from plants. Therefore, it would appear to be to our advantage to keep something growing on the land throughout the entire growing season. It will benefit all life if we capture all of the energy possible and keep the soil life working twenty-four hours a day throughout the entire season.

In summary, we can say that farming involves managing the entire soil-plant system. When we discuss growing crops we should also be able to say that we grow soil life because crop production at its best is concerned with both.

The management of soil fertility will be covered in depth in the Soil Fertility sections of this book. Management of soil air, water and decay directly affect soil fertility and the recovery of nutrients by the plant. Air is necessary for soil life, root proliferation and the uptake of nutrients and water. Too much or too little water can adversely affect nutrient availability and soil life.

We strongly believe that the areas of air, water and decay management are mostly influenced by tillage practices or the physical management of the soil. Chemistry also plays a role in soil structure, decay needs, etc. While chemistry is important tillage practices are the most influential.

For these reasons tillage practices may be our most important "system" to design and manage because of its tremendous influence on air, water and decay. Let us remember that soil fertility is defined in terms of both soil nutrient levels and the ability of the plant to recover those nutrients. Because tillage practices can either favor or adversely affect nutrient release and uptake by the plant we must give top priority to tillage principles and practices.

TILLAGE PRINCIPLES & SYSTEMS

Fertility management is the final act in our farm show. The first three acts in the drama were the management of soil aeration, soil water and residue decay to set the stage for the last act, management of soil fertility.

We place fertility management last because we cannot have a fertile soil unless those first three acts have been done well. A soil's fertility is judged by the response of the crop to that fertility. A crop's response and growth is based on the balance of soil fertility and its recovery by the plant. High fertility is of little value unless it can be efficiently recovered by the crop.

We have learned the value of soil oxygen for root growth and in nutrient uptake by the roots and recognize the relationship between water management and soil fertility that makes it necessary to avoid either water excesses or deficiencies. We have a basic understanding of the many benefits bestowed on the soil-plant system by the decay process.

We will now briefly interrupt our study of the four basics of soil management and place the study of tillage ahead of the fourth basic, soil fertility. This is done because of our belief that tillage has the greatest impact on the first three basics, soil air, soil water and decay, and is a prime prerequisite of fertility management.

Tillage affects soil aeration by its positioning of crop residue which influences its decaying process. Correct decay will impart a biological granulation of soil particles and assure soil aeration. Deeply buried residue will not decay and will deprive the soil of this natural granulation process. A barren soil surface seals over and impedes the exchange of soil gasses. Properly positioned residue has a positive effect on soil aeration by promoting decay and preventing surface sealing.

Tillage also influences water management by positioning residue to either encourage or discourage water insoak. It is tillage that creates barriers such as plow pans and it is tillage that can remove them. Tillage can either permit or prevent water insoak and penetration to storage, thus affecting its recovery by the plant.

For these reasons we believe that the tillage stage must be properly set before working on the final act, soil fertility. The final act will "make" or "break" the show, so all the props must be in place before it begins.

Development of an effective tillage system requires careful coordination of two basic areas. First, we must carefully define soil management principles. In other words, we must know precisely the tillage requirements necessary to meet the needs of the soil-plant system.

Second, we must select the tools that are properly engineered to meet these carefully defined soil management requirements.

Tillage may be considered a science in the respect that it requires the knowledge and integration of the soil and plant sciences to design systems that, when used correctly, will meet the needs of each.

The primary purposes of tillage are to enhance the growing crop and, at the same time, minimize damage to the soil system. All tillage is detrimental to the soil to one degree or another, but it must be considered as a necessary evil in the growing of crops. This may come as a surprise to some, so let us briefly review the ways in which tillage damages soils.

Tillage speeds the oxidation and decomposition of organic matter by aeration of the soil. From the moment the virgin prairies were first plowed there began a slow but steady decrease in humus. The original levels of twelve to fifteen percent or even higher of humus have often been reduced to as low as four percent or less. The early sod-busters would not recognize their ash-bed soils if they could see them today. Under natural conditions soils carefully regulated their oxygen levels and were never over-exposed as they are through tillage.

Nature regulated soil aeration very carefully through residue management on the surface and also throughout the naturally firm soil horizon. The surface residue not only stopped erosion, but prevented the soil surface from sealing over. The living and dead roots provided the "wick" action needed to permit air to penetrate or exchange through an unsealed surface. This principle must be considered in the design of tillage tools and systems.

Tillage also causes rapid drying of soil which creates hardened clods and greatly influences the microbiological systems. Over-aeration, erosion, destruction of soil texture and biological fluctuations are sufficient to impress upon us that tillage is, to a degree, soil destructive.

Erosion may be the major cause of soil destruction. We believe that proper tillage is necessary for high yields on most soils. We must recognize, however, that these high yields can also be accompanied by massive soil erosion. Therefore our goal must be to obtain high yields for an indefinite length of time by managing tillage in a manner to make certain the soil-building processes exceed the soil destruction.

Therefore **each tillage system must be specifically planned to enhance the management of soil air, soil water and residue decay.** However, proper design and proper use are not always coordinated at the farm level. The best designed tillage tools can be misused and fail totally in their originally planned purpose. An understanding of the soil basics is essential in the effective application of properly engineered tillage tools.

Success in the implementation of a tillage system begins with selection of well designed equipment. There are a staggering number of different tillage implements available in the market place. A farmer must be able to understand his specific needs and to recognize the best equipment to fulfill those needs. In this respect the farmer and the design engineer must both be well schooled in the soil basics. The most popular tools are not always the right ones, with the moldboard plow being a prime example of this.

Often tools are designed to salvage mistakes and to relate to a single operation instead of a segment of a planned system. The tillage industry has been a "buyer beware" game where farmers have misplaced their trust and served as guinea pigs in the testing of agricultural iron. Much of this iron has ended as "barnyard furniture," sitting idly around the farm.

After having reviewed the soil-forming processes, become acquainted with the soil-plant relationship and of the plant requirements, we are still not quite ready to go shopping for a tillage system. We still must understand soil characteristics and establish goals for primary and secondary tillage. Finally, our knowledge must culminate in the development of a tillage system for our particular farm and area. From now on, when we go to the field to till we must go there with real purpose.

Soil compaction is a primary concern when working with the physical management of most soils. The term "compaction" implies a pushing together or compression of something which resulted from an application of force or weight. Gravity is the force that gives weight to an object; therefore we can conclude that gravity is the villain causing compaction in our soils. Gravity is the force that tries to pull everything to the center of the earth.

When looking at a wheel track, as in Figure 8, we see the effect of gravity trying to pull the tractor to the center of the earth. Gravity pulled the tractor down a few inches until the resistance of the soil overcame its pull. This resistance under the wheel caused compaction, which is our number one problem. Note the compaction pattern as it wedges out in width below the ground.

Any time soil is compacted it is at the expense of soil air. Compaction forces soil particles more closely together, displacing soil air simply because the pore spaces become filled with soil particles. It is the same as filling the spaces with water except that water will eventually be gone.

The volume of air required to fill the wheel track indentation is the amount of air that has been pushed from the soil. The degree and depth of compaction will depend on the weight of the equipment and the texture and moisture level of the soil.

Fine-textured soils with high moisture content will compact most easily. A moisture film on soil particles serves as a lubricant and reduces friction. Moist soil particles slide easily, thus increasing both depth and intensity of compaction.

Dry soil particles have more resistance or friction which will reduce depth of compaction. The moisture lubrication concept is important because of its impact on the depth and severity of compaction. This should impress upon us all that being first in the field in the spring may be a very shallow victory. The wheel tracks, shown in Figure 10, will take a toll on roots and will not be removed by spring tillage or cultivation.

Increased soil density with the resulting loss of soil air as shown in Figure 8 is usually sufficient to stop roots from entering the compaction zone. Figure 6 shows how a wheel track will turn roots downward. No matter how well the surface is refitted the lower compaction remains. The surface refitting, or seedbed preparation, imparts additional compaction at the lower levels. Once the importance of soil air is understood every field operation will be carefully appraised. **We must think of a "tillage system" in terms of one that will meet the soil management requirements with a minimum of trips over the field.**

As we begin the process of selection of a tillage system we will first review some past practices, starting with the moldboard plow which is still the dominant primary tillage tool in use today.

We will look at the plow objectively and with no adverse reflection upon those who are still using it. Since the plow is still the primary tool in use, those who plow are in the majority. Remember, however, the majority is not necessarily always right.

MOLDBOARD PLOWING

Figure 39 illustrates the principles of the moldboard plow. We begin with the fall plowing phase and carry it through the spring seedbed preparation.

The moldboard plow, as shown in the "fall phase," turns the soil upside down and completely inverts it, leaving the surface bare and free of protective residue. The individual raindrops, acting as fluid bombs, pelt the soil surface causing separation of the soil particles and sealing of the soil surface. This will guarantee runoff, making the first goal of water management, getting the water to soak in where it falls, impossible to accomplish.

A smooth, sealed surface or freeze-dried layer also creates a wind erosion condition. The force of wind loosens soil particles permit-

THE MOLDBOARD PLOW SYSTEM

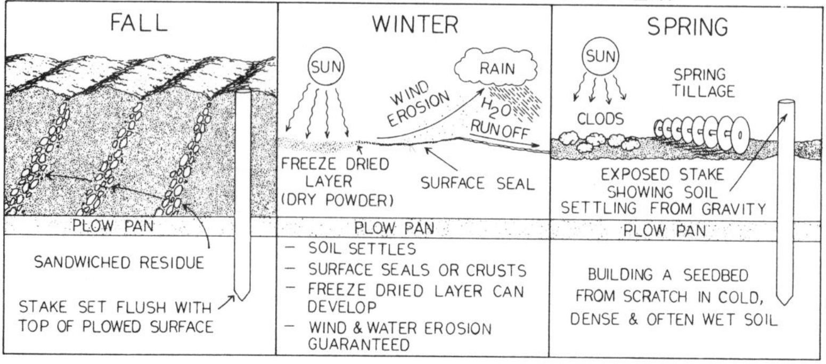

Fig. 39

ting them to roll across the surface and break loose other soil particles. Freeze-drying results in a thin layer of powder-dry soil. Bare, frozen soil can absorb the sun's rays and cause surface thawing to a depth of one-eighth to one-fourth of an inch. The moisture in this thawed zone can quickly evaporate into the cold, dry winter air leaving a thin layer of dry soil at the surface. This layer will remain powder-like and dry because moisture cannot move up from the frozen layer below. These conditions can bring about dust storms and cause severe soil losses.

This is illustrated in the "winter phase" of Figure 39. In effect, nature says the soil surface should be covered and protected. To this extent nature has a plant species that will grow and cover the soil surface under nearly every conceivable soil condition.

Plowing in the fall guarantees that the vehicles of erosion, water and wind, will be set into motion. The plow also sandwiches the residue causing several other problems while depriving the plants of the benefits of decay. Often residue is plowed up in the spring that was plowed under the previous season simply because the plow has placed it below the depth where a fence post can decay. Thus we lose the benefits of decay and the release of the inorganic and organic nutrients.

Decay at lower depths is of an unfavorable type that can release alcohols, aldehydes and formaldehye, which further preserves the buried residue and also creates a toxic condition for roots. The sandwiched residue can also interfere with the downward movement of water and break the capillary rise back to the surface. The burying of residue obviously creates a multitude of problems. We have covered these in the section of this book on **Managing Decay.**

On many soils the plow also imparts a plow pan. The plow works at a depth where the clay concentration is high. As the plow slides through moist clay it smears the clay particles (illustrated earlier in Figure 21), in an action similar to that of a trowel over fresh cement. It smears, seals and compresses the soil into a compacted zone or plow pan that can extend more than an inch in depth.

A plow pan will effectively separate the plow zone from the lower soil and limit the depth we can farm. Roots cannot penetrate the dense zone. Water movement into storage and capillary recovery back into the root zone will be greatly impaired. This separation nullifies any efforts to manage water and places the farmer at the mercy of either too much or too little water.

Most soils contain a topsoil and a subsoil zone. It may be compared to a two-layered cake. Figure 22 illustrates a wheat problem and how a plow pan limits farming to plow depth. Removing the plow pan, as illustrated earlier in Figure 25, permits farming of both layers, or the entire "cake."

The capillary rise of water (recovery), (shown in Figure 28), cannot function if residue or a compaction barrier is placed between the subsoil and the root zones.

Figure 22 with its explanation of a wheat problem in Ontario pretty well summarizes the problem of soil barriers.

The plow lifts the soil against the force of gravity leaving the surface higher in elevation from where it was. If a stake is set, as shown in Figure 39, we will find the stake sticking well above the surface by the following spring. Again, this is due to the force of gravity trying to pull the loosened soil back toward the center of the earth.

In Figure 39 we can see that by spring the soil has been beaten down by rain and the pull of gravity into a compacted condition. Keep in mind that soil compaction always results in the loss of soil air. Dense soil is lower in air and usually higher in moisture than loose soils. Since water requires more heat to raise its temperature than an equal volume of soil or air compacted soils may be slower to warm up in the spring.

Spring arrives and it is time to go into the field to prepare a seedbed. By force of habit we are thinking in terms of building a seedbed from scratch with iron. We enter the field, as shown in the "Spring Phase" of Figure 39, with an assortment of iron, with cabin fever at its peak, well ahead of the neighbor, but happily creating real problems for ourselves. It is at this point that we may be about to initiate our largest loss of potential yield, especially if we have medium to heavy-textured soils.

In the spring conditions are most ideal for deep compaction and the formation of clods. Figure 40 shows how clods are formed when a compressed and smeared volume of wet soil dries out too quickly.

Fig. 40

Wet clay particles have mushy surfaces that smear together forming a weak concrete, but a pretty solid dirt clod, upon drying.

Clods are formed from our most productive topsoil and are non-productive for that season. The formation of clods can result in salvage trips across the field in the attempt to build the seedbed. We have coined a phrase that says, "we can sin in the fall, but never in the spring." In other words, most soils can be worked in the fall even if it is very wet because wetting, drying, freezing and thawing will help to mellow the soil before spring.

However, if we "sin" by working the soils too wet in the spring we can create a situation such as illustrated in Figure 40. It will require extra trips to salvage this field. These extra trips to manicure the surface cost money in power and time and take a fearful toll causing underground compaction and loss of potential yield.

The art of using iron in its various forms to mold a seedbed from flat, dense and often water-logged soil has never been fully mastered.

The persistence of this system has built into our farming programs a guarantee of increased costs and reduced yields. A tillage system must impart compatability between operations. The fall phase should complement spring tillage by reducing its hazards.

As we learn more about soil management principles we realize that the plow is not a system and does very little to complement operations or soil processes. In fact, there is very little that can be said in favor of the moldboard plow. This does not mean that it has no place under certain conditions, but certainly not as a widely-used tool or practice.

It would be ironic if history recorded that the downfall of western civilization was brought about by the plow! When we consider that soil erosion losses are probably greater today than they were even thirty years ago, we cannot call it "progress."

With all of its shortcomings we have not yet seen a mass movement to replace the plow. Any major tillage change directly affects the livelihood of the farmer and he should not change until he is totally convinced there is a better way. While the argument for erosion control is a valid one, it has not been sufficient by itself to create a major shift away from the plow. A farmer will grudgingly sacrifice his soils before he will venture into a major change in tillage that may threaten his living. This is a hard reality that must be recognized. In the past too many tillage systems have failed because both the manufacturers and the users were not certain of what they were attempting to accomplish.

In general, tillage operations may be classified under three major headings, **subsoiling, primary** and **secondary** tillage practices. Each individual tillage practice must be a part of a well-planned system designed to manage the soil basics of air, water and residue decay. **Our major tillage goals must be to develop an ideal seedbed and to improve the soil-plant marriage throughout the growing season.**

General soil management principles apply to all soils, but the means of accomplishing these principles will vary. For instance, air management can be a serious problem on heavy clay, but no problem at all on muck, coarse sand or gravel.

Wherever possible primary tillage should be done in the fall on medium to heavy-textured soils. There are certain soils that need only secondary seedbed tillage due to their ease of management. Heavier soils are hazardous and can be damaged very easily when worked in the spring. On these soils we must make every effort to minimize spring trips.

Terminology is important and should eliminate confusion instead of causing it. Let us not call a spade a "shovel" even if they are interchangeable. The terms, "subsoiling," "primary tillage," "secondary tillage" and "no-till" can very nicely cover all of the operations which make up our "tillage systems."

Because we consider "no-till" a practice rather than a system it will be treated as a separate entity. The terms, "subsoiling," "primary and secondary tillage," must be thoroughly understood before an effective tillage system can be developed.

SUBSOILING

The common term "chiseling" should not be confused with subsoiling. Let us think of the subsoiler as a tool to be used in the removal of plow or clay pans. A subsoiling operation is shown in Figure 24 and its purpose illustrated in Figure 25.

A chisel plow, as such, should be considered a primary tillage tool designed to thoroughly incorporate crop residue and fertility throughout the topsoil layer or root zone. On the other hand, the subsoiler is designed to break through pans and clay layers with a minimum of soil mixing. We want to avoid bringing the lower clay or low-organic subsoil up to the surface.

A standard tillage system will include **subsoiling** where needed and **primary** and **secondary** operations. Each phase has its individual goals and purposes leading to a total system that is complementary to both the soil and the plant. Each phase must be studied both individually and collectively. Methods and purposes of subsoiling are explained in the unit **Managing Soil Water.**

PRIMARY TILLAGE

Primary tillage may be defined as any major tillage operation that properly sets the stage for the secondary, or seedbed preparation, phase. Certain soil types do not require primary tillage to enhance the secondary tillage stage. Where it is needed, however, primary tillage should be a fall operation designed to fulfill the following objectives:

1. Residue should be thoroughly incorporated within the biotic zone of decay, leaving enough on and sticking through the surface for maximum water insoak, aeration and to minimize erosion, as illustrated in Figures 41 and 42.
2. The operation should position the soil into modified ridges which is also shown in Figures 41 and 42.
3. The entire operation, except the removal of pans, should be done in one operation if at all possible.

In addition to these three objectives, any plow, disc or clay pans that are present should be removed, as outlined in the section on **Water Management** and shown in Figure 25.

Fig. 41

Fig. 42

MODIFIED RIDGE SYSTEM

The modified ridge system is one method of primary tillage and predicated on storing soil in a ridged position to be leveled prior to planting. We call this a "modified ridge system" as compared to the "ridge planting system" which will be outlined as an alternative to consider.

In the modified ridging system the residue is incorporated throughout a broad-based ridge that ends up from eight to ten inches in height on thirty-inch spacings. This is accomplished with twisted-shank tillage tools, such as the Landoll Soil Master. This tool is illustrated in Figures 43 and 44 and is often referred to as a "disc chisel."

Fig. 43

Many equipment companies are presently manufacturing this type of tool. The modified ridging with twisted shank tillage is rapidly replacing the moldboard plow and proving to be the most positive change in tillage since the introduction of the moldboard plow.

The twisted-shank tillage system is designed to meet the three primary goals of residue management, ridging and a one-trip operation. If understood and used properly these tools are extremely effective in a very wide range of soil conditions.

However, to be effective these tools must be used properly. They come equipped with a gang of cutting coulters up front, as shown in Figure 45. These coulters are spaced in seven and one-half inch intervals and designed to run directly in front and between each shank. Their purpose is to cut residue into specific lengths to prevent plugging, but more important, so they will serve as pipettes or wicks to assist in the penetration of water. They are designed to work on

Fig. 44

Fig. 45

solid ground with no previous tillage. We believe the straight cutter blades are superior to coned-disc blades that tend to overwork the soil ahead of the twisted incorporating blades.

These tools are designed for a one-trip operation, so working the soil ahead of this operation should be avoided unless absolutely necessary. Sometimes residue is so heavy that discing is necessary ahead of ridging. Stalk chopping should be avoided because it does not increase or speed decay; instead it destroys the wick action and can cover the surface with small pieces of light-colored stalks that reflect sunlight and slow the warming process in the spring.

Figure 46 shows the four-inch-wide, single piece, reversible blade that is very popular and long-wearing. The four-inch blade lifts and turns more soil than narrower blades. This is important in building a ridge and properly incorporating the residue.

Fig. 46

Figure 47 shows the soil action off of the twisted shank. The shanks work like miniature moldboard plows in the way they lift and turn the soil.

Fig. 47

Depth is also a consideration. These tools are designed to go no deeper than the plow. If they went deeper they would penetrate the plow pan or high-clay area and roll it up to the surface. As mentioned earlier, we do not want to bring the high-clay layers to the surface. In other words, these tools are not subsoilers.

Speed may be the most important element in the successful use of these twisted-shank tillage tools and they need to be pulled at five to six miles an hour if they are to work correctly. If pulled more slowly the soil action comes lazily off of the blades and results in poor incorporation and ridging. Speed is important and the tool must be designed to pull at these speeds without bucking or porpoising through the soil. Care must be taken to buy a tool with good design and engineering.

If you cannot pull these tools over five miles per hour you will be unhappy with the results. Figure 48 shows the results of pulling a tool too slowly on the right side and at the proper speed on the left. Notice the difference in residue incorporation and ridges. The picture illustrates the difference between four and five and one-half miles per hour.

Speed is more important than depth because without adequate speed these tools will not perform regardless of depth. At five to six

Fig. 48

miles per hour the soil is carried up the entire length of the blade and curled off the top to meet the soil from the opposing blade. It is this action, shown in Figure 47, that properly incorporates the residue and builds the ridge.

Figure 49 is a diagram showing the entire system. The fall operation thoroughly incorporates the residue and builds the ridges which end up thirty inches from peak to peak. These ridges minimize water

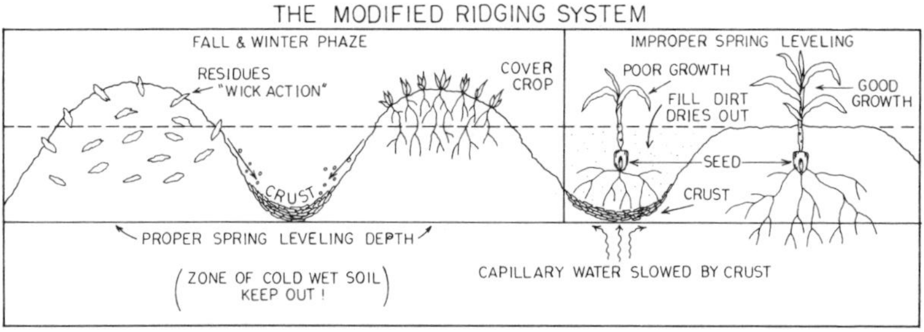

Fig. 49

runoff even on four to six percent slopes. The ridges divide up the slope into a series of miniature watersheds. This prevents the accumulation of water from a broad area coalescing into a rampaging stream and causing severe gullies, as shown in Figure 50. During excessive rainfall the only water that runs off is the water between the two ridges that cannot soak into the soil.

Fig. 50

On steep slopes the tool can be run across the slope on a contour to buffer runoff. These passes can be spaced every one hundred to three hundred feet, depending on length and severity of slope. Figure 51 shows how ridges at right angles across the watershed can serve as coffer dams in slowing down the water and permitting the soil to settle.

The residue, properly cut into lengths by the front coulters and incorporated throughout the ridge, will decompose because it is within

Fig. 51

the biotic zone "where a fence post will rot." The residue serves as wicks to get the water to penetrate into the surface where it falls instead of running off. This is important in water management and well illustrated in Figure 42.

The ridges also prevent wind erosion because the winds cannot get the clean sweep over the surface that would begin the process of loosening and rolling soil particles along the surface to create a sandblasting effect which could culminate in a dust storm. The swirling air over a ridged field acts as a dry cleaner pulling soil out of the air. In this manner you may be able to trap soil blowing in from the surrounding fields. Ridges help stop the two vehicles of erosion — **water and wind.**

One of the ridges in Figure 49 shows a cover crop. If a cover crop such as rye is to be planted it should be seeded ahead of the primary, twisted-shank tillage. When the proper speed is maintained the seed of the cover crop tends to float upward and ends up on top of the ridge as shown in Figure 52.

Fig. 52

Some farmers blend their rye or wheat seed with their fall fertilizer and broadcast the combination of fertilizer with cover crop in one trip ahead of primary fall tillage. Others use a small, electric motor-driven hopper mounted on the front of the tractor to spread the cover crop while performing the ridging operation. The main purpose is to avoid an extra trip across the field whenever possible.

Primary tillage must set the stage for managing soil air, soil water and residue. This in turn will minimize spring weather-related hazards in the secondary, or seedbed preparation phase. As outlined, the modified ridging system is very effective in reaching these objectives, particularly on medium to heavy-textured soils.

SECONDARY TILLAGE (SEEDBED PREPARATION)

Figure 49 also shows the spring phase of the modified ridge tillage system. Storing the soil over the winter in properly built ridges is the most practical method of minimizing spring weather hazards. Rays of the sun quickly warm up the ridges creating an early environment for soil organisms to begin their important task of decaying residue and granulating soil particles.

A ridge can be from five to fifteen degrees warmer than the flat soil next to it. Ridges have better air and water control than flat soil. As stated earlier, it requires far less heat to warm a well-aerated soil than an equal volume of water-saturated soil.

Ridges mellow out and develop a seedbed on their own. In the spring we have often noticed how much more mellow the top of a back furrow will be in comparison with the lower, flatter soil on each side. The development of a seedbed without tillage tools is possible with ridges. It is more profitable and less hazardous to begin spring tillage in a naturally pre-formed seedbed than it is to build a seedbed from scratch with iron!

The purpose of ridging is to store the soil properly so that it will develop a seedbed prior to spring tillage, the objective being to level seedbed material in the secondary tillage phase instead of building a seedbed in flat, cold and dense soil.

Several things must be considered when preparing the seedbed from the modified ridging system. For instance, the depth the ridges are worked is important. The thickness of the crust that forms in the valley between the ridges must be checked. Soil particles wash off the ridges and settle in the valleys, forming a crust up to one inch in thickness.

This crust is shown in Figure 49. When working or leveling the ridges while preparing the seedbed the tillage tool should be set to work only as deep as the crust in the valley between the ridges and no deeper. While the ridges are mellow the soil beneath them can be cold and wet. If the spring tool is worked into this zone it will smear the cold, wet soil and bring it to the surface to dry out into clods. This is a good way to damage the seedbed.

Figure 53 clearly shows the results of working only as deep as the base of the ridge. The farmer in Figure 53 is using a light disc with a harrow pulled behind. He has installed gauge wheels so that he can precisely control the depth of the disc. The result is an ideal seedbed.

Fig. 53

The right portion of Figure 49 shows the wrong way to level the modified ridge. Serious problems can develop if the ridges are leveled only part of the way down. Partial leveling fills the valleys with loose soil and does not disturb the crust formed at the base of the valley. This crust can slow the capillary rise of water from below. As a result, the fill soil in the valleys can dry out.

The soil under the ridge will be firm and remain moist by capillary movement from below. Seed placed in the valley area can be slow to germinate while the seed in the ridge area will germinate and grow normally. This condition can cause very uneven growth, as shown in the right side of Figure 49. For these reasons it is very important to work the ridges to the correct depth while preparing the seedbed.

Another important point to realize is that the ridges are a means of reducing spring hazards. Ridges store the soil in a protected position and should be leveled just ahead of the planting operation. Ridges will permit a farmer to go to the field well ahead of the neighbor who is preparing a seedbed from a flat surface. But, leveling the ridges too soon could set a farmer back to the position of his neighbor because then, he too may have to rebuild a seedbed from flat, run-together soil. Ridges must be maintained right up to planting time. **Don't lose the advantage.**

Figure 42 shows proper fall ridging with the stalks cut into short lengths, poking into and out of the soil and acting as wicks to regulate water insoak and aeration. Figure 41 is an example of the modified ridging process at its best.

Once soil management principles are understood this type of tillage takes on an appearance of beauty. These modified ridges prevent wind erosion and greatly reduce water erosion. The residue is positioned where it will decay and the soil warms up faster and preforms a seedbed as a result of microbiological activity and air and water management. These soils will soak up water where it falls.

As beautiful as they look, these ridges are not complete if we permit the existence of a plow pan that prevents the insoaking water from penetrating down to storage. It is a real tragedy to have a ridged field more wet in the spring than a neighbor's plowed field because a plow pan was not removed. And if soils turn out wetter than the plowed fields, they could also have more weeds.

Our primary tillage goals have now been established and ridge-leveling procedures for preparation of the spring seedbed outlined. Primary and secondary tillage must be thought of as a system and never as isolated operations. The goals of primary tillage are designed to enhance the secondary, or seedbed preparation, process.

The spring seedbed or secondary tillage operations have their own objectives. Our "primary tillage" practices and goals were designed to help consistently achieve the secondary, or seedbed, objectives. We must emphasize that seedbed preparation is not the sole purpose of a tillage system, but a single entity in the total soil management process. This concept is important because it influences other decisions about tillage, such as "no-till."

Certain principles of secondary tillage must be understood for selection of proper spring tillage tools. When going to the field in the spring, what do we want our secondary tillage equipment to accomplish? Machinery that fulfills the following main objectives will make an ideal seedbed for most crops.

1. The tillage tool must provide proper chemical incorporation and seedbed preparation in a single trip to reduce time, cost and compaction.
2. It must provide satisfactory results on flat or ridged surfaces.
3. It must manage residue by breaking it into smaller pieces and depositing the coarser material on the surface to prevent surface crusting or sealing.
4. It must structure the seedbed by sorting of soil particles, depositing the finer particles below and the coarser material above the seed, to prevent clods and air pockets from breaking the rise of capillary water up to the germinating seed.
5. It must have the capacity to operate at high speeds with low maintenance.

There will be exceptions with crops such as beets and small seeded vegetables that must be placed at depths of one inch or less. For such crops the producer will probably demand a residue-free surface. His seedbed goals may be completely opposite to what we normally consider to be ideal. However, regardless of the individual situation we must always go to the field with an exactly defined purpose and know what we are attempting to accomplish with a tillage operation.

After having implemented primary tillage we are ready for the selection of the secondary, or fitting tools, that will best meet our seedbed requirements. This may be difficult because of our quest for perfection and the tremendous number of tools available from which to choose.

But, since we know the spring seedbed preparation goals we can quickly discard many of the tools as being unworkable. With the "modified ridge" system in place there are more tool options for spring tillage. Ridges simplify the task of preparing a seedbed in a one-trip operation.

Some farmers select the field cultivator, as shown in Figure 54, or a light disc, such as shown in Figure 53. Notice that the disc in Figure 53 has gauge wheels to precisely control the depth to the base of the ridge. It also has harrow sections mounted behind. The field cultivator in Figure 54 could profit from a rear-mounted harrow that would break up many of the clods showing behind this tool. One disadvantage with both the field cultivator and the disc is that it will probably require two trips across the field to properly incorporate herbicides. More than one trip must be avoided if at all possible.

Some tools, such as Landoll's "TILLOLL," have been specifically engineered to meet the five seedbed goals. See Figures 56 and 57. The TILLOLL represents a new generation of spring tillage tools that will do it all in a single trip with high ground speeds and low upkeep.

Fig. 54

Many tools are presently available and there will be new designs constantly being developed. It is a seller's world and everyone would like to sell a piece of iron. Much "barnyard furniture" is scattered across rural America as a monument to engineering failures and superior salesmanship.

With understanding of soil management requirements we can recognize poor engineering and resist the best iron salesman. Tillage

Fig. 56

Fig. 57

then evolves into a system instead of single, unrelated practices. Each farmer must have the knowledge and capability to put together a system for his own conditions.

Figures 24, 58 and 59 represent a tillage system. The V-plow is a subsoiler designed to remove the plow or clay pans shown in Figure

Fig. 58

Fig. 59

25. This permits water to penetrate into storage and return when needed. It permits roots to penetrate deeper. It builds continuity between the topsoil and the subsoil so that the entire "cake" can be farmed and not just the top layer.

The twisted-shank tillage operation shown in Figure 58 meets primary tillage goals and the "TILLOLL" shown in Figure 59 meets the five seedbed management objectives. Such a system is right today and will be right in the future because it is right with the soil.

Because one system will not fit all soils every farmer must know the basic soil requirements and then adjust his system to fit his particular situation. He must make his tool selections guided with knowledge and experience. The system shown in Figures 24, 58 and 59 will fit quite a wide variety of soil conditions.

In summary we conclude this portion of our study of tillage with a few thought-provoking statements.

1. The term "conservation tillage" may be, in the same category as "organic foods" or "organic farming" — poor terminology. It implies the selection of a tillage system solely for the purpose of soil conservation. Soil conservation is only one aspect of a total soil management system. All tillage practices should conserve the soil, and if a practice does not, it simply should be abandoned.

   The moldboard plow does not conserve soil so it should be abandoned as a general practice. A tillage system must have well-defined objectives such as those outlined under "Tillage Principles and Systems."

   Our endeavors to manage soil air, water and residue decomposition are prime factors toward increased yields. When these areas are managed **soil conservation is automatically taken care of.**

   It is much easier to bring about tillage changes when we can show the logic of total soil management. "Conservation tillage" will not sell, except to the few totally dedicated, because it is just too confining. All farmers would like to stop erosion, but most will grudgingly sacrifice their soils rather than change practices that have been long established. It is simple human nature to look after one's self-interests. Conservation tillage may be compared to "tunnel-vision" — it does not address the whole picture, despite the good intentions behind it. If we manage the basics soil conservation will be accomplished without making it the central issue.
2. There is a general belief that intensive farming and high yields are in some way causing massive soil destruction. This is simply not true. High yields or intensive farming by itself does not

cause the destruction of soils. Instead, it is the individual practices used in production that cause soil losses. The simple truth is that total soil management will increase the quality and quantity of yields and, at the same time, conserve the soil. The practices that cause soil erosion will eventually put severe limitations on both yields and quality. We can obtain high yields while destroying our soils but not for very long.

3. Tillage is always soil destructive. It is a necessary evil in assuring crop survival and production. Tillage is a one-sided act; it favors the plant at the expense of the soil. This is why it is so important to gain an in-depth knowledge of soil science so that we may till the soil with minimal damage. The necessary soil destruction brought about by farming must not exceed the ability of the soil to mend itself. This must become a primary objective of everyone who tills the soil. With this objective achieved we will be able to farm the soil until the next glacial period!
4. Proper soil tillage results in the best management of soil air, soil water and the decomposition of residue. These three areas determine the availability of nutrients and their uptake or recovery by the roots. **The effect of a good tillage system can, under certain conditions, equal or even exceed the effect of adding fertilizers to the soil.** For these reasons the tillage system is an extremely important area of management. This is not to deny the importance of maintaining soil fertility. It simply means that we must think beyond the simple act of adding fertilizers. Because the fertility of a soil is determined by the crop's ability to recover nutrients anything that can be done to improve recovery will increase its fertility. Fertility, after all, is worthless unless it can be recovered by the plant.
5. Soil losses from prime farmland are more severe today than at any period of time during the last twenty or thirty years. These increasing losses continue in spite of the Soil Conservation Service's efforts and dedication during the last three to four decades. The vast majority of prime farmland has little or no conservation practices in use. Soil losses are so severe that legislation may become necessary to curb it. We do not need more legislation; we need education.

Figures 60 and 61 are prime examples of the problem. The purpose of these waterways appears to be solely to protect a gulley! It will not be long until these waterways will be filled with silt from the totally unprotected surrounding field.

Tax dollars were used to build this waterway. Contrast this with Figure 62 which shows a tillage system that properly utilizes residue to manage soil air, water and residue decay. The difference between

Fig. 60

Fig. 61

these two farmers is education. One understands soil management principles and the other does not.

The Soil Conservation Service and supporting agencies, in spite of their good intentions and farm planning, must depend upon the farmer to implement and maintain recommended practices. Why do we continue practices such as those shown in Figure 60? Why would a land owner allow such massive soil destruction as that shown in Figure 63?

There may be many reasons why we either do not start or else discontinue conservation practices. Inconvenience may be a factor. We may feel more comfortable doing the same as others around us. We may fear change because of past experience or the experiences of others. Certainly we have the right to fear failure. We may be overly sensitive to peer pressure. We may subconsciously have come to accept soil losses as part of the cost of farming. The economic and

Fig. 62

Fig. 63

production problems of today may have dimmed our concerns for the future. Whatever the reasons, we are failing at conservation.

This failure will not go unnoticed or unchallenged. A change toward strict conservation is inevitable. Change must either come voluntarily or it may be forced through legislation. It is our decision and not much time is remaining in which to make it.

In addition to the modified ridging system there are other alternative practices and systems that may be considered. No-till and ridge planting with all of its variables are becoming quite popular. The system ultimately adopted must be successful. Success depends upon our understanding and meeting the soil management principles of air, water, residue decay and fertility. Any system that can manage these basics will succeed.

We will continue our unit on tillage by studying the Ridge-planting System, No-till, the practice of cultivating row crops and end row management.

RIDGE PLANTING SYSTEMS

Ridge, or bed planting, is an alternative system of tillage that may have orginated in Chinese agriculture, dating as far back as three thousand years. Ridge planting is a common practice in many parts of the United States, particularly in the west and southwest. The midwest has been slow in adopting the practice and as a result may be missing an opportunity for increased yields and efficiency. Midwest farmers who have adapted to the practice appear to be very pleased with the results and often express their opinion that they will never return to conventional flat land farming.

Ridge planting, as considered here, refers to the system of planting directly on top of a ridge that has been prebuilt prior to planting. It must not be confused with the system that levels the ridge during the planting operation and then rebuilds it by hilling unconsolidated materials back into the row while cultivating. The major differences between these two systems will be outlined in this unit.

Ridge planting may well be the ultimate system with respect to many of our midwest soils. Bed planting has been used for the obvious reasons of facilitating flood irrigation, stopping wind erosion and as a means of managing poorly drained soils. In addition to these there are many other advantages of planting on a ridge.

1. Ridge planting permits the farming and utilization of more topsoil than conventional flat land farming. The topsoil is the most productive soil strata and may be viewed as the "action zone." With conventional tillage the benefits of two to three inches of topsoil may be lost because of clodding and drying. Furthermore, we impart wheel track compaction to large volumes of topsoil between or under the rows. With the ridges the inter-row topsoil is gathered and heaped into a bed so that the less productive subsoil is compacted between the rows and the loss of topsoil through clodding, drying and compaction is minimal in relationship to total topsoil volume. Farming more of the topsoil has always been our goal and this becomes a reality with ridge planting.

2. Ridges simplify the management of soil air and water. Excessive water in heavy textured or poorly drained soils can virtually eliminate soil air during critical periods of plant growth. The primary goal in water management is to maintain a full thickness of capillary film and to prevent the filling or flooding of the air spaces in the soil. On most medium to heavy-textured soils this goal cannot be totally or consistently managed when the soil is maintained in a flat position. Unless totally covered with water, a ridge prevents the flooding of airspaces and maintains a very desirable capillary influence.

Some people are concerned that a ridge may be more prone to drying out than flat soil. In effect, the opposite is true. When properly constructed a ridge controls soil water and will often maintain a capillary film long after flat soil reaches the wilting point in moisture levels. Properly stored water from below tends to move upward in a wick or capillary action to the **highest soil level.** A good capillary rise of water into the ridge and root zone can be maintained as long as there is a supply below. The moisture rising into the ridge must pass through the root mass before reaching the surface. The management goals of getting water to soak in where it falls, penetrate into storage and recover back into the root zone applies regardless of the tillage system being used. Ridges greatly improve the entire water management process.

The main advantage of the ridge as it relates to water control is that it controls the wild and damaging fluctuations in soil moisture levels often encountered on flat positioned soils. Uniformity in water management leads to a more uniform total soil environment.

Soil microbiological activity is more intense, thus creating improved soil structure and aeration. Flooding of air spaces is virtually eliminated along with its disastrous effects on soil and plant life. How many times during an average year have we experienced periods of total reversal of ideal soil conditions for plant growth because of too much water? These reversals rob growth time and are damaging to plant nutrition by causing a deficiency of soil oxygen. The ridge system is the "equalizer" for the soil-plant system. Soil air and water control are the two most important aspects of soil management and are controlled best with ridges.

3. Ridges permit the development of larger and more efficient root systems. This is only natural since we have eliminated wheel tracks under the row, provided more topsoil for the root mass and a more consistent and uniform water supply along with soil oxygen, which are the keys to root growth and nutrient uptake. These all add up to a more consistent and proficient recovery system.

A large root system is very important, but it must also function properly throughout the life of the plant. A large root system may become worthless if flooding eliminates soil air.

Figures 64 and 65 show excellent examples of how a ridge can improve the root system and yields. On a heavy Canadian soil ridges improved soil aeration to permit an extensive root system to grow under these harvester tomato plants. The more extensive root system, as shown in Figure 65, greatly improved nutrient uptake and as a result produced the excellent crop of tomatoes shown in Figure 64. The ridged tomatoes on these tight, heavy-textured soils produced twelve to fifteen more tons of tomatoes per acre than those planted

Fig. 64

Fig. 65

on the flat. This farmer produced those extra tons of tomatoes with added soil oxygen. Soil aeration to provide oxygen for root growth and nutrient recovery is the cheapest source of fertilizer we have.

4. With good air and water management we also stabilize the soil microbiological environment. This, in turn, enhances and speeds the proper decomposition of residue and the subsequent release of nutrients and other biological factors necessary to the well being of the plant. Proper aerobic decomposition is very important. Anerobic decay organisms that thrive in the absence of soil oxygen can cause gasification losses of nitrogen and sulfur.

5. Ridges maximize soil fertility. The fertility of a soil is judged by the ability of the plant to recover its nutrients. Soil oxygen is necessary because of its role in root growth and nutrient uptake by the roots. By stabilizing soil air, water, and thoroughly incorporating fertility throughout the ridge the plant can more efficiently utilize soil fertility and, in effect, increase the fertility of the soil. Everything works better in a ridge.

6. The ridge system controls wheel track compaction. **There are no compaction tracks under the row.** Also, when the topsoil is gathered into a ridge we become less concerned about wheel tracks between the rows. The purpose of the ridge is to utilize as much topsoil as possible to build an environment where roots can go deep under the row.

7. Herbicide use can be reduced by twenty-five to forty percent with ridges. There is no problem with cultivating ridges. In fact, secondary tillage and cultivation are both done with a rolling cultivator. Of main concern is the use of an herbicide band directly over the top of the row,

8. Secondary tillage to prepare the seedbed is done with the rolling cultivator which requires less power than discs, field cultivators and other spring tools. This operation also has the advantage of operating at high speeds. On a corn-soybean rotation ridge planting may eliminate primary and possibly secondary tillage on soybean ground going to corn which may substantially reduce costs.

9. Finally, we must emphasize the tremendous importance of decaying the bulk of corp residue **within** the soil system rather than on the **surface.** The purposes of surface residue must be to prevent surface sealing and erosion by wind and water. Ideally, the surface residue should be kept to a minimum to accomplish these goals with the balance of residue properly incorporated to decay within the soil system.

Surface decomposition of residue is a penalty we must pay with no-till to minimize erosion when we insist on row cropping severely sloping land. Surface decay contributes little, if any, humus to the soil system. Carbonaceous residue on the soil surface decomposes and oxidizes releasing the carbon into the atmosphere as carbon dioxide instead of becoming humus. Medium to heavy-textured soils may become increasingly dense with no-till.

With the ridge-plant system, as illustrated in Figure 65A, the ridges are completely leveled and rebuilt in the fall which thoroughly incorporates the carbonaceous residue and fertility throughout the entire ridge. The decomposition of the residue will benefit the entire soil mass.

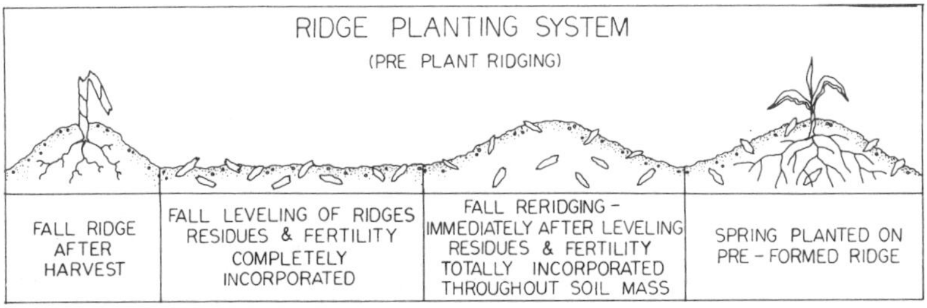

Fig. 65A

Because decomposing residue is blended throughout the soil, the succeeding crops will also have improved soil conditions and reap the benefits caused by increased humus, improved soil structure and improved water supply and aeration. These factors lead to better root systems, soil fertility, nutrient recovery and a progressive improvement in quantity and quality of production.

Ridge planting offers many advantages over flat land farming and will gain in popularity as farmers begin to understand the relationship between ridging and the ease of control over the soil basics of air, water, decay and fertility.

Let us now review a farm situation where crops are successfully grown by planting them directly on top of a pre-built ridge. The Olthoff brothers, Bernie and Don, of Grant Park Illinois, have been using the ridge-plant system since 1969.

The sequence of operations used by the Olthoffs is illustrated in Figures 66, 67, 68, 69, 70 and 71. They use a corn and soybean rotation. In the fall they level the beds on the corn fields with a heavy offset disc. The soybean fields are not leveled.

Immediately following the discing operation the Olthoffs use their John Deere integral bedder with markers, as shown in Figure 66, to re-establish their beds on thirty-inch spacings. The bedder originally had double discs on each spool. For thirty-inch row spacings the smaller disc blade can be removed. The double disc, however, should be used for thirty-six inch or wider row spacings.

Fig. 66

Fig. 67

Fig. 68

Fig. 69

Fig. 70

After the bedding operation the ridges will be about ten inches high as shown by Don in Figure 72. In Figure 73 note the excellent incorporation of residue which has been doubly mixed to decompose throughout the soil mass instead of just on the surface.

Fig. 71

Fig. 72

Fig. 73

For obvious reasons the end rows are left level as shown in Figure 74. End row ridges could be laid up in the fall at right angles to the field ridges and leveled in the spring. It is better to store the soil in ridges over the winter, especially on heavier-textured soils.

Fig. 74

Fall work is greatly simplified once ridges have been established. However, when laying up beds for the **first time** extra measures may have to be taken to assure adequate residue incorporation and enough loose soil to build a proper bed. The pre-ridging operations will differ depending upon the previous crop and soil conditions. A good sequence for heavy cornstalk ground may include discing as deeply as possible, followed by a disc chisel with a drag or reel attached to the rear to level the ridges formed by the twisted shanks.

Regardless of the number of trips required, it is very important to build a proper ridge the first time. The pre-ridge tillage must provide enough loose workable soil to assure the incorporation of residue. It would be a mistake to sandwich a mat of unincorporated residue into the center of the beds.

We prefer to establish the original beds in the fall. If they are laid up in the spring careful pre-ridge tillage becomes even more important. Ridging in the spring requires more precise control of soil conditions than fall ridging and, if conditions are not right, waiting until the following fall. An unconsolidated and poorly constructed ridge could dry out and become a disaster.

As spring arrives the beds may appear as shown in Figure 75. Notice that the tops of the beds are starting to warm up and mellow into a seedbed even though the valleys and lower portion of the beds are still frozen.

The seedbed preparation and incorporation of herbicides is accomplished with a rolling cultivator as shown in Figures 67, 68 and 69. We believe that a rolling cultivator is a "must" for ridge planting.

The Olthoffs bank a rolling spider gang on each side of the bed and a four-roller spider directly on top of the bed. An herbicide with forty to sixty pounds of liquid nitrogen is applied and **thoroughly** incorporated in the ridge during the same operation. The balance of their nitrogen can be applied during the first cultivation or through the planter.

The rolling cultivator reshapes the bed and prepares an excellent seedbed. It will require one or two trips across the field, depending upon soil conditions. The rolling cultivator is the only spring tool needed and provides a high speed, low cost and low power operation.

Planting immediately follows after the seedbed has been prepared with the rolling cultivator. The planter, as shown in Figure 70, is blocked in the "up" position and stabilized by two rigid coulters mounted on the front end of the planter. This can be seen in Figure 71. A typical seed placement is shown in Figure 76.

The same rolling cultivator, minus the top spider, is used to cultivate both the corn and soybeans. The spider gangs can be adjusted on the sides of the beds to work as aggressively as necessary.

Fig. 75

Fig. 76

The Olthoffs apply the balance of their nitrogen needs while cultivating. Nitrogen, plus some 10-34-0 and anything else that may be needed, is laid in a stream close to the plant and covered in the hilling process. Yes, you can hill corn on a ridge with a rolling cultivator and you will be able to appreciate the high efficiency of nitrogen usage in a ridge after you study **Nitrogen Management** in the Soil Fertility section of this book.

Figures 77 and 78 show the excellent results the Olthoffs consistently experience with corn and soybeans planted on ridges.

When corn follows soybeans there is little need for fall, or primary, tillage. In the spring the old soybean bed can be reshaped with the rolling cultivator and corn planted directly into the old bean row. The old bean roots cause no problem and actually improve the root zone for the following corn crop. The four-spider gang run on top of the bed will break off most of the old bean stubs and prepare an excellent seedbed with herbicide incorporation. A corn-soybean rotation on ridges would reduce primary tillage by fifty percent and represent considerable savings.

Some growers, as shown in Figure 79, do not believe it is necessary to do any tillage ahead of planting on old soybean ridges. This would make an ideal no-till situation in comparison with the usual no-till on flat land.

As a third crop the Olthoffs grow onion sets. They plant two rows on top of each bed, as illustrated in Figure 80. It should be noted how uniform the onions are across this rolling field. The onion sets are grown from seeds so they require a fine seedbed at planting time for precise seed placement. For this operation the Olthoffs flatten the tops of their beds, as shown in Figures 81 and 82.

Figure 83 shows poorly drained land that can benefit from ridging. Ridging can delay or eliminate the need for expensive tiling in heavy-textured and marginally drained soils. Bernie and Don Olthoff are prime examples of two farmers who have successfully mastered the ridge planting system.

Howard Keys, from Beason Illinois, shown in Figure 79, is another progressive farmer with a successful ridging operation.

Planting on ridges is the surest method of gaining growth time through the combination of earlier planting and the minimizing of weather-related setbacks during the growing season. Ridge planting fits many soil conditions and is especially effective in areas with short growing seasons or on soils that are heavy-textured or poorly drained. Ridges permit farming of much more topsoil than is possible with conventional farming.

Fig. 77

Fig. 78

Fig. 79

Fig. 80

Fig. 81

Fig. 82

Fig. 83

POST-PLANT RIDGING

If a tillage or planting system does not complement the soil basics it will not endure. Planting systems are being promoted that offer modified versions of the pre-plant ridge system.

A typical system is illustrated in Figure **83-A.** Under this system the previous year's ridges would be removed or lowered during the planting operation and then rebuilt with one or two hilling operations while cultivating. This system rebuilds the ridge after the crop is growing instead of building it before planting. These two systems can be referred to as "pre-plant" and "post-plant" ridging.

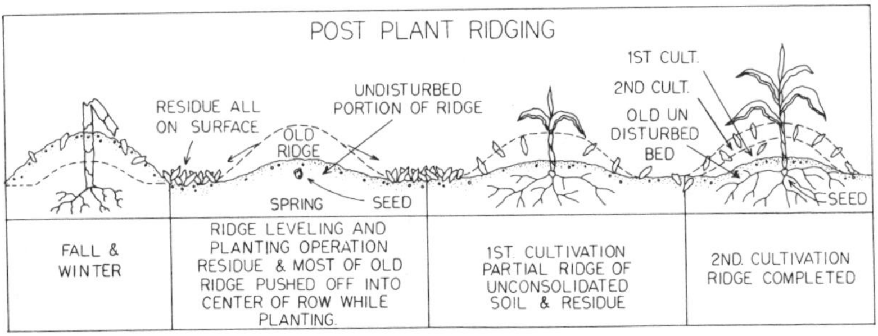

Fig. 83A

With the post-plant ridging method secondary tillage is combined with the planting operation. As with no-till this can generate attention because of its simplicity and newness. However, the primary purpose of tillage is to complement the four basics of **soil air, soil water, residue decay** and **fertility.** Any tillage tool or system that consistently fails to enhance one or more of these basics will not endure. This concept cannot be overemphasized.

Because so many new types of tools and innovative approaches to farming are being promoted we have formulated the following two concepts to serve as guidelines while evaluating a tool or system.

Number one, the bulk of crop residue should decompose within the soil system. The only time the bulk of carbonaceous residue, such as cornstalks and small grain straw, should be permitted to decompose on the surface is when we farm no-till on sloping land where there is no other alternative for minimizing soil erosion.

This statement correctly implies that soil erosion must be stopped at all costs. Once the soils have been lost to erosion we will no longer need to concern ourselves with "soil basics." The future must not be sacrificed for short term gains in the present. Our soils must endure for generations and we are convinced that decomposing crop residue within the soil system is essential for this longevity. If we fail to stabilize or build soil humus the management of soil air, soil water and fertility on the medium to heavier-textured soils will become increasingly difficult.

And, **number two,** the author believes pre-plant bedding is usually superior to post-plant bedding. If we plan to utilize ridges, why not go all the way and plant directly on top of a pre-built ridge, as illustrated in Figure 65A?

Planting directly on a pre-formed ridge is superior to the concept illustrated in Figure 83A. It shows clearing the residue and partially leveling the ridge in front of the planter, placing the seed in the freshly exposed soil before it can warm up and then rebuilding the ridge by rolling unconsolidated soil and residue back into the row with one or two cultivations.

Post-plant bedding, as illustrated in Figure 83A does not properly manage fertility or residue. With this system the seed is planted in the same undisturbed soil year after year as in no-till. The ridge is lowered by the same amount each spring during the planting operation. This part of the ridge is pushed off to the center of the rows and then moved back during cultivation. Two cultivations to rebuild the ridge are done with a very wide sweep run between the rows. It would appear that this kind of tillage between the rows could be damaging to the roots.

It is doubtful that the material used to reform the ridge during cultivation would have time to consolidate or firm up enough to support roots. And even if it did, the late decay of residue around the plant would certainly compete with the plant for nutrients. Also, broadcast fertilizers cannot become incorporated within the root mass under the row. The fertility is constantly being moved out and then back into the row and must remain unavailable to the plant for much of the season. By the time the bed becomes firm again it will be time to push it off with the planter to start the whole cycle over.

This system is similar to no-till because it does not incorporate fertility within the root zone and most of the residue decomposes or oxidizes on the surface instead of within the soil mass that supports roots and production.

We must be concerned that the post-plant ridging, as in Figure 83A, may fail in the management of soil air and fertility in many soil types. This is also a major concern with no-till.

However, post-plant ridging may be considered as an alternative to no-till on soils where the surface management of residue is necessary for the prevention of erosion. The system maintains most of the residue on the surface and places the seed in a better seedbed than that which is possible with straight no-till. Management of fertility would still be a major concern.

In summary, we should again stress the importance of decaying the bulk of residue within the soil system rather than on top. Humus and soil-building potential should not be sacrificed because all of the residue has been left to decompose and oxidize on the surface unless it is absolutely necessary for the prevention of soil erosion.

We must make every effort to farm all of the topsoil possible, so why go to the effort of moving the best topsoil off the ridge and then moving it back into the row with two root-pruning cultivations? It would appear that we are actually trying to keep fertility and the best topsoil out of production and expending a lot of energy in the process. By planting directly on top of a pre-plant ridge we will not have to sacrifice or compromise any of the soil basics.

No-till and post-plant ridging practices can be useful where all of the surface residue is needed to prevent soil erosion. Utilization of meadow crops instead of row cropping would be our choice if we could always adhere to sound land-use practices on severely erosion-prone soils. This would simplify tillage and good land use would minimize the need for systems such as no-till with the confusion and frustrations of attempting to adapt them to situations where they do not fit. Understanding the four soil basics and the refusal to compromise them will make the selection of a tillage system simple.

WEATHERPROOFING CROP PRODUCTION

The ultimate in tillage systems would be one that made it impossible to damage the soil during seedbed preparation. Ridge or bed planting, whereby crops are planted directly on top of beds laid up prior to planting, probably comes as close to the ultimate in this respect as any system presently in use. As a result ridge planting offers the best potential for weatherproofing our crops.

Weather is the least controllable of crop production factors. We are most concerned with rainfall and temperature, with rainfall obviously being the most important and also the most unpredictable. Excessive water can cause as much harm as too little — both can destroy or reduce yields.

The ideal production management system would protect or buffer the crop from either excesses or deficiencies of water. The goal is to maintain a sufficient capillary supply of water under both extremes.

To accomplish this we must develop a two-tiered root and fertility system. This is outlined in Figure 83B. Planting directly on top of a properly constructed bed is in itself a major step toward weatherproofing the crop.

As outlined earlier, the bed provides maximum uniformity in relation to soil aeration and water supply within the soil environment. As a result of air and water control we will have maximum residue decay, root development and recovery of fertility. The beds give excellent protection against excessive water and improve water movement into the root zone. They have a directional effect on the movement of water from the lower supply upward toward the peak of the ridge. This pattern of flow must pass through and become exposed to the root mass on its upward journey.

The ridge and its mass of roots provide the very best "zone of updraft," as depicted in Figure 83C. This zone of updraft concept is very significant to a plant's growth as explained in the unit on **Managing Soil Fertility.**

The bed planting system provides the most consistent control of the four soil basics — aeration, water, residue decay and fertility. In this respect it is probably the ideal system to use where it can be adapted. The ridge planting system can be additionally improved if a second tier of roots and a fertility base can be established in the lower soil stratus directly below the bed. This can often be accom-

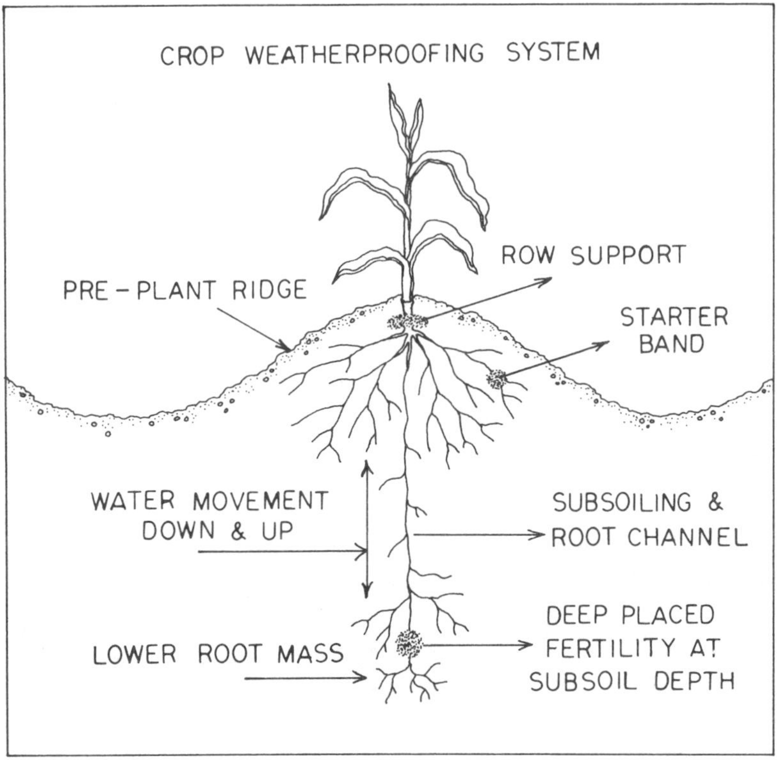

Fig. 83B

plished at the time of bedding with a narrow chisel tine set to penetrate six to twelve inches below the base of the ridge.

An injection of a fertilizer blend at the bottom of the chisel slot would also enhance the weatherproofing process. The correct fertilizer blend to use would depend on the needs of each particular soil. Phosphorous, along with some nitrogen, either as a liquid or dry blend (13-52-0, 18-46-0 or 10-34-0) would be very important for stimulation of root growth at this depth. A study of the **Soil Fertility** sections of this book will help you determine your own fertility needs. Soil testing at this lower depth may become necessary.

Under normal conditions the crop could benefit from both soil zones. If the lower soil became saturated the plant's roots would thrive in the beds. If the weather turned extremely dry and dried out the beds the lower roots could still function.

Weatherproofing of crops is a new concept, but it could prove to be the best insurance we have and should be pursued with vigor.

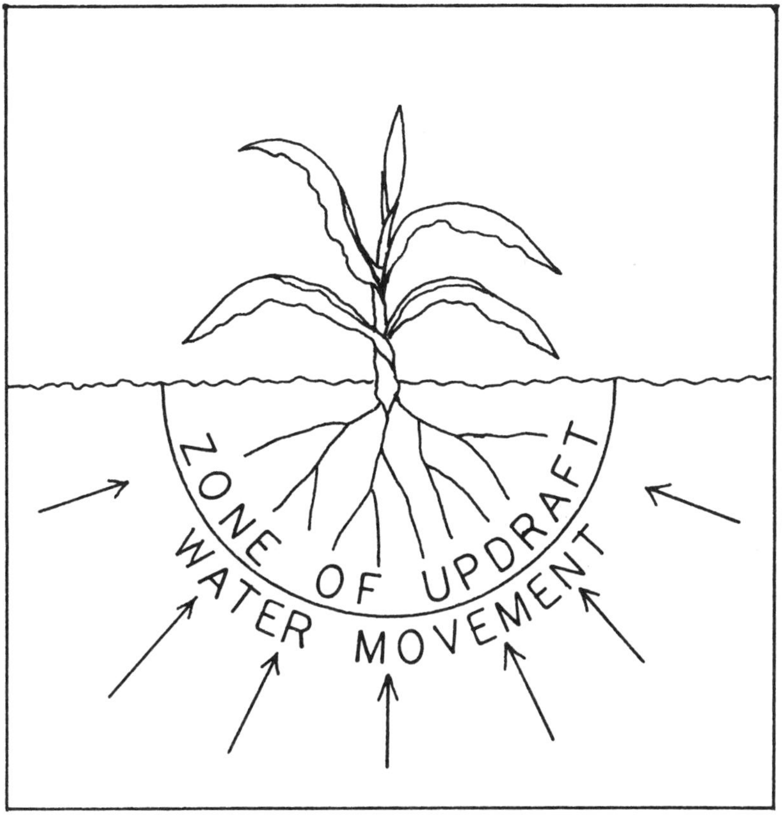

Fig. 83C

NO-TILL

We cannot discuss tillage without making reference to the practice of "no-till." No-till has been around for a long time. But despite its many advocates it has not gained wide support throughout the cornbelt states.

No-till, in the author's opinion, is a practice and not a system. It is a practice based on the concept of reducing costs and erosion. These are goals toward which we all should strive. Planting without working the soil has the potential to reduce erosion just as fewer trips over the field will reduce costs.

However, a tillage system must solve problems and not create them. You must carefully appraise your own soil situation and ask yourself these questions to determine whether or not no-till will work for you.

1. Will it adequately manage my soil aeration?
2. Will I be able to manage soil fertility?
3. Can I control soil warmup in the spring?
4. Will my seedbed be adequate for good germination and a healthy start for my crop?
5. Will it manage water and residue decay?
6. Will I be able to control weeds and insects?
7. Will it maintain or increase yields and profits?
8. Will yields and profits remain consistent over a period of years?

Any tillage system should answer these questions in the affirmative. It seems doubtful that the practice of no-till will consistently accomplish these goals for the reason that it is being recommended too generally.

Where, then, does the practice of no-till fit and work best and consistently? We believe no-till fits under the following conditions:

1. It fits extremely sloping land that cannot be conserved with conventional tillage or cannot be adapted to normal conservation practices such as contouring and/or strip cropping. No-till is the only alternative when we insist on row cropping land that should not be planted to crops requiring tillage. This type of land should be in pasture, hay or even timber. Millions of acres of such land should be legally off limits to all tillage equipment, except possibly for no-till, and then under prescribed guidelines. The practices that lead to land abandonment after the topsoil has eroded away should be outlawed for the good of all. Figure 84 illustrates successful cropping with no-till on land normally too hazardous to row crop.

2. No-till can fit the practice of double-cropping, such as beans following wheat. Moisture can be a limiting factor for the second crop. Tillage can dry out the soil and delay germination sufficiently to jeopardize maturity. Moisture is the primary concern and is best

Fig. 84

conserved with no-till. Excessive residue, when incorporated, can also interfere with a quality seedbed and set up nutrient competition from decay organisms. When left standing small grain stubble will reflect sunlight and keep the soil cooler which will reduce surface evaporation.

3. Climate, heat degree days and length of growing season greatly determine the success of no-till. As we move **south** of an east-west line through central Illinois no-till becomes increasingly more feasible and successful. As we move **north** the practice of no-till becomes more and more hazardous and difficult to manage as a general practice. Soils with residue-laden surfaces that are dense, low in air and high in moisture warm up very slowly in comparison with soils worked with tillage systems designed to manage soil air, water and residue decay. Further north, no-till can delay planting and set the stage for serious weather-related setbacks, making it more difficult to manage a proper seedbed to assure the all-important processes of germination and early seedling development.

4. Corn or other row crops, following beans, offers an opportunity for use of no-till. Bean ground is prone to erosion due to the loose soil structure and minimal residue. For this reason fall tillage is risky because it buries too much residue leaving the surface quite barren. It may be best to simply leave all of the bean residue uniformly scattered over the surface as shown in Figure 85, to break the impact of rainfall and reduce erosion from runoff. When spring arrives you will notice that the old bean row has developed a beautiful seedbed without tillage. The seedbed in the old row can be additionally improved if you practice moderate hilling when cultivating the beans.

Fig. 85

Figure 86 clearly shows the nearly ideal soil conditions in the old bean row. This offers an excellent opportunity to practice no-till.

Fig. 86

An actual operation of no-till corn being planted in old bean rows is shown in Figures 87 and 88. Gaylord Lottinville and sons Kyle and Kevin in Sheldon, Illinois have their planter set up for no-till and row support (Figure 87). They plant most of their acreage with conventional tillage, but take advantage of no-till planting where it fits. In Figure 88, Kyle is showing the adverse soil and weather conditions experienced into June throughout most of the midwest during the spring of 1983. In this heavy soil no-tilling corn into the old bean rows was not ideal, but under the poor spring conditions it was far superior to building a seedbed with conventional tillage.

A no-till coulter will prepare an excellent seedbed in the old bean rows, or a regular disc opener by itself will be enough, particularly if it was hilled a little during the last year's cultivation. The old bean root system loosens the soil and provides organic channels for the roots of the next crop to follow. Planting no-till in the old bean rows may be the most ideal situation to incorporate no-till as part of a full tillage system.

No-till must not be implemented without a complete appraisal of the situation. This demands an in-depth understanding of soil management principles as they relate to the chemical, physical and biological properties of the soil.

Fig. 87

Fig. 88

Those who sell the no-till equipment and principles must understand the soil and plant system's needs. Selling no-till as a means of increasing chemical or iron sales is wrong. No-till has its place and it is our responsibility to see that it is used properly and effectively. It is also the farmer's responsibility to sufficiently understand soil management principles so that he can wisely select a system or practice to fit his situation.

As a general practice no-till has its greatest potential in the southerly latitudes. In the north the potential for no-till decreases rapidly. Soil types become a major deciding factor whether it can work consistently and drainage becomes critical. Well-drained sands, gravels, or high organic soils may not benefit from primary tillage or deep chiseling. These soils can, and possibly should, be farmed with only secondary or no-till practices.

In the areas of shorter growing seasons we need to adopt practices that will provide the maximum use of growth time. Anything that may take away or reduce the total time allotted for a crop to grow and mature must be avoided.

The "start period" for a crop is very important. Much potential yield is lost in the very beginning. A crop's potential is set and determined early in its life cycle. The number of rows of corn on a cob is set during the ovulation period that takes place about five weeks after the corn emerges. The number of kernels per row is set about nine weeks after emergence. These are critical periods for establishing the potential yield.

Yield potential for many other crops is set early in their growth cycles also. Therefore we must do everything possible to assure a good beginning. Practices that adversely affect the start will reduce genetic potential as well as total growth time. The combination can guarantee reduced profits and possibly even failure.

Each general area of the country has yield-limiting factors even with top management practices. Management practices should be designed to try to get the most from the season by eliminating or minimizing hazards. It is unwise to drastically alter tillage practices without a thorough review of their total impact on the soil-plant system.

The shorter the growing season, the more important soil management becomes. In the areas with shorter seasons the best way to gain growth time is to assure early warmup and the ability to maintain good soil conditions on earlier plantings. This is not easy to do. The best method may be ridge planting, which was discussed at length in the preceding unit. The second-best method may be the modified ridging system also outlined earlier. As we move north into heavier soils and shorter growing seasons no-till becomes increasingly hazardous and difficult to manage. A cold soil cannot merely be opened up and the seed dropped into its adverse environment with consistent results. It is certainly not the way to gain growth time and minimize hazards.

Fertilizer management is extremely difficult under no-till. Medium to heavier-textured soils may become more dense and root restricting with no-till. The reduced soil aeration that limits root growth also limits nutrient recovery. Many farmers fail to build high fertility levels throughout the top seven inches of soil prior to starting no-till.

Surface broadcasting of fertilizers can be very inefficient. Figure 89 shows how fertility levels can vary at different depths in an actual no-till situation on medium to heavy-textured soils. Note the extreme differences between the top three inches and the four inches immediately below. These samples were taken from the same soil probe. It clearly illustrates the problem of topdressing fertilizers.

The extreme variations in phosphorous and potassium levels would also apply to surface-applied limestone and trace elements. With the exception of mobile nutrients such as nitrates and sulphates most nutrients are less mobile and will not penetrate much deeper than one to two inches. Surface-applied nutrients are generally ineffective because of a lack of uniform moisture and the inability of the

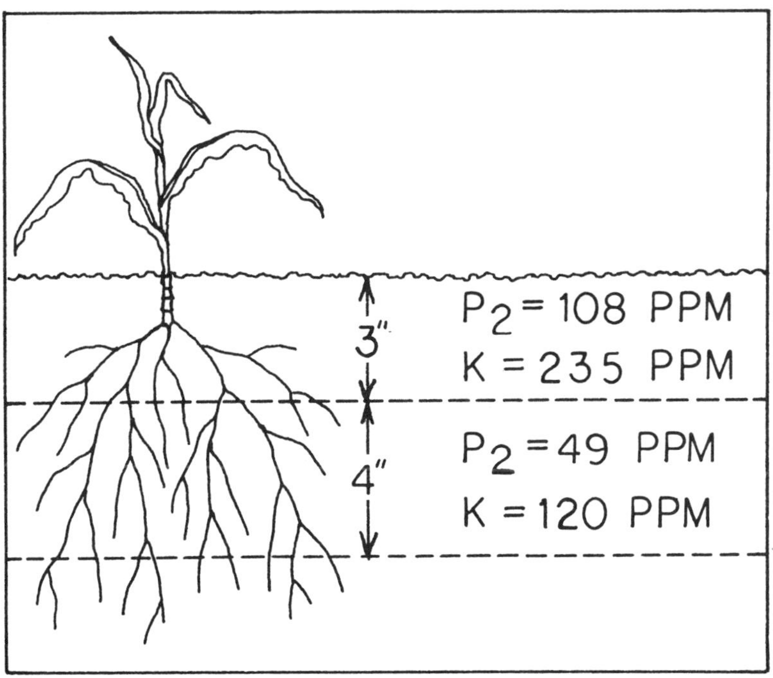

Fig. 89

nutrients to penetrate into the root mass. Root growth and nutrient uptake ceases whenever the soil moisture level approaches the permanent wilting point.

The upper two to three inches of soil is very vulnerable to drying. Roots remain ineffective in dry soil until moisture conditions improve and permit new root hairs to develop. Soil moisture and aeration are both necessary for nutrient recovery.

On medium or heavy-textured soils poor aeration at the lower depths and surface dryout can combine to severely limit production under no-till, and even if aeration and moisture are adequate at the lower depths, fertility could be lacking.

To get an accurate fertility reading the lower four to five inches of soil should be tested separately from the upper two to three inches. It is possible, as shown in Figure 89, to have a very high soil test reading if the upper three inches of soil are included in the sample. This could mask a low reading below despite the high average reading.

Fertility accessibility and availability are the challenges with no-till and soil aeration and nutrient placement become prime management factors. Our intention is not to totally discourage the practice of no-till, but rather to emphasize its hazards and pitfalls.

Seasons and weather patterns are unpredictable and represent a major area of risk. Most cultural practices are directed toward minimizing the hazards of weather. All farmers should farm as though they were in a shorter season than they actually may be and this would place emphasis on giving the crops a good start.

We know the hazards of planting in a cold and wet soil and what poor drainage and tight soils can do to root function and yields. We also recognize that we must do whatever is necessary to grow a crop. Sometimes this may mean two or more tillage trips. Saving trips across the field is an excellent idea providing it is not at the expense of the crop.

In reality the practice of no-till has a wide variety of applications and we may conclude that it will work in many instances providing we are able to manage the three soil basics of air, water and fertility. Notice that the fourth basic, residue decomposition, was excluded because all of the residue is left on the surface with no-till.

We must remind ourselves that surface decay does not build soil humus. No-till as a regular practice will gradually deplete internal humus and ultimately cause most soils to become increasingly dense and compacted.

We urge that no-till be used especially on severely sloping land where normal tillage will trigger severe erosion. If these same erosion-prone soils were planted to non-tillable crops such as hay or meadow we could use no-till where it is best suited, such as double-cropping beans in wheat or planting in old bean rows. **In this respect a no-till planter would be a good investment because it would do an excellent job with conventional tillage, but permit saving of tillage under the right conditions.**

The no-till operator must realize that basically he will be farming within a box, as illustrated in Figure 90, that measures five to seven inches deep and fourteen to sixteen inches wide. This is the "action zone" that will contain most of the crop's roots. Fertility and aeration must be managed within this box. If the basics of soil management cannot be handled within this box of soil no-till will fail. We have designed a Soil Fertility-Soil Aeration System for this action zone and it is thoroughly covered in the special section of this book entitled "Fertility Management for No-Till."

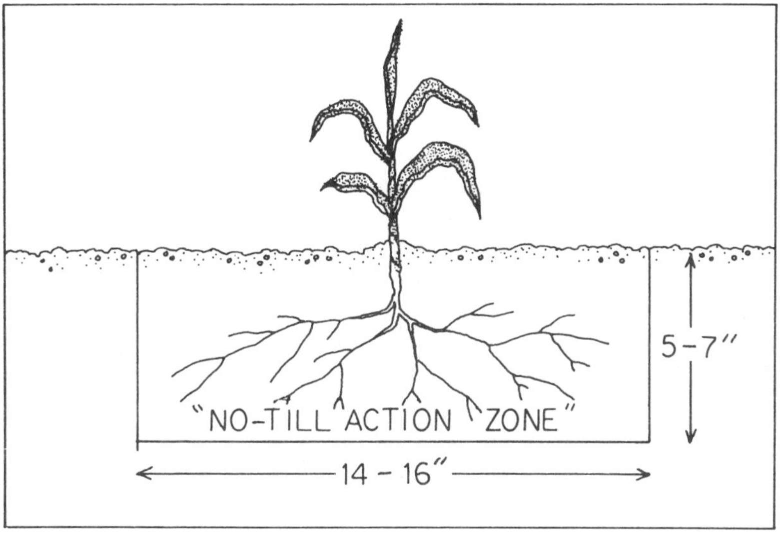

Fig. 90

PRINCIPLES AND GUIDELINES FOR CULTIVATING ROW CROPS

When herbicides were introduced during the 1950's many farmers began to abandon the practice of cultivation. As confidence and experience in the use of herbicides grew, more and more farmers became dependent upon them as a means of controlling weeds. The abandonment of the cultivator with the advent of herbicides indicates a belief that weed control was considered the primary function of cultivation.

Studies supported herbicides by showing that the practice of cultivation did not generally increase corn yields providing weeds were controlled. This data obviously encouraged industry to invest heavily in the potentially lucrative field of herbicides. Considering the massive effort of the herbicide industry and the data indicating cultivation had no purpose beyond weed control, the lowly cultivator became an endangered species. For many, abandoning the cultivator meant one less chore and it was appealing to simply put away the planter and await harvest time.

By the late 1970's we were generally experiencing a renewed interest in cultivation in many parts of the country. In fact, many had never completely stopped cultivating. The return of cultivation may be triggered by the high cost of herbicides and the general fear of chemicals or possibly because farmers are realizing that there may be more to cultivation than simply the control of weeds.

We should not conclude that the abandonment of cultivation was a wise decision. The continued use or discontinuance of a practice does not necessarily reflect the true value of it. For instance, it would be a mistake to approve of plowing simply because many farmers are doing it.

We view cultivation as a tillage practice. Should it be considered in broader terms as an essential part of a total tillage system? We can justify cultivation providing we can show that it benefits the growing crop.

Some farmers will state their belief that it is just good business to stir the soil around a growing plant. In this respect we have often noticed the stimulating effect cultivation can have on a crop such as corn. Often within a day or two after cultivation the corn will appear to improve in color, increase its rate of growth and in general take on a flush of renewed vigor. **These are sure signs that this particular soil needed aeration.**

From our study of soil aeration we learned the importance of good aeration and the dire consequences resulting from its absence.

Improving soil aeration may be the greatest benefit derived from cultivation. The uptake of nutrients, root vigor and microbiological activity, as we learned earlier, are oxygen-demanding processes.

It is becoming more difficult to manage soil aeration on the medium and heavier-textured soils. Many of these soils are much tighter today than they were twenty-five years ago. The loss of topsoil and the depletion of humus has increased the soil density, making air management more difficult. When we decided years ago that cultivation did not pay we were working with more friable, humus-laden soils that required less physical management to maintain aeration. We must face the fact that much of this has been lost during the years. Farmers who lived at the turn of the century would not believe their eyes if they could farm the same soils today.

We can be certain that cultivation will pay on any soil where oxygen becomes the limiting factor. How do we determine if a soil is deficient in oxygen? Asking what a soil's potential yield is may be a better question. On most medium and heavier-textured soils oxygen deficiency can be a major production limiting factor, and we suggest that if you farm these types of soils you consider making cultivation a permanent part of your total tillage system. The need for soil aeration can be the best reason for owning and operating a cultivator.

In addition to managing soil aeration there are three important functions that should be built into every cultivator:

1. A cultivator should, for certain soil conditions, have the ability to penetrate deeply between the rows.
2. A cultivator should have hilling capability.
3. A cultivator should be able to place a liquid or dry nutrient close to the row while cultivating.

DEEP PENETRATION

The primary purpose of deep penetration capability for a cultivator is to assist and improve soil aeration. The process of "**gas diffusion**" as a means by which soils breathe is an important process that can be enhanced through deep tillage between the rows. During periods of excessive rainfall, certain soils become very dense and airless. The density may extend throughout the root zone and actually cause the crop to stop growing. Under these conditions a normal cultivator may not penetrate sufficiently to correct the problem. A narrow shank, such as an anhydrous knife, penetrating five to six inches in depth between the rows under these conditions can bring about a dramatic improvement in plant growth and vigor.

Several years ago an experiment of injecting propane gas between corn rows with an anhydrous applicator caused much excitement because it increased yields. The excitement over injecting propane faded when it was discovered that yields were just as good where the knives were run through the soil without propane! The deep penetration permitted the **lateral diffusion** of carbon dioxide and oxygen in and out of the root zone. Improved soil aeration, not the propane, caused the better yields.

In tight, anerobic soil conditions deep tillage between the rows can save the day. Deep tillage too near the plant, as illustrated in Figure 91, would destroy too many roots. However, when the soil is tight and vigor reduced, pruning a few roots in the center of the row is a small price to pay for the potential gain.

The capability to deep-till between the rows should be an option that is used only when needed. If the soil is adequately aerated there is no advantage to tilling deep between the rows. Under certain soil conditions deep tillage may over-aerate and cause drying which would become detrimental rather than helpful. Certain soils would never

Fig. 91

benefit from the practice. But, deep tillage could pay and become a standard practice on many soils. Consider it a rescue or salvage operation and use it accordingly. When in doubt set up tests and check yield differences.

HILLING

The second feature a cultivator should have is the capability to place a hill around the plant. View the hilling process as growing the crop on a miniature ridge, as outlined under the **Ridge Planting System,** the difference being that the ridge is laid up after, rather than before, planting.

A ridge, three inches high, can help manage both air and water around the base of the plant. The condition of the soil which is placed in the hill is important. Placing clods or excessive trash around the plant would be of little advantage. The soil should be in a condition that will permit it to firm up so that roots may establish themselves within the ridge.

Certain root and lower stem rots in beans are associated with tight, anerobic soil conditions around the plant. Soil aeration is necessary for the prevention and stopping of these diseases. A modest two to three-inch hill around a bean plant can prevent sealing and crusting conditions from developing.

Hilling can permit corn to develop one or two extra sets of nodular roots. A node that is covered will shoot an extra set of underground roots. The brace roots that extend vertically from an aboveground node, as shown in Figure 91A, develop a massive set of fibrous roots if they can penetrate the soil surface. The brace roots as shown in Figure 92 should have been covered when they first appeared as small bumps at the base of the plant when the corn was twelve to fifteen inches high.

These upper roots are important to corn because it appears they are mostly responsible for feeding phosphorous to the upper part of the plant late in the season. Good uptake of phosphorous is important for maturity, grain fill and weight in corn. In turn this could reduce drying costs at harvest time. Upper root development is simply one of many factors involved in producing a profitable corn crop. Hilling can help in this process.

Hilling can also be beneficial in covering and controlling weeds in the row. Crops such as corn can be hilled more extensively than beans. Occasionally weeds get out of control in corn and require a salvage program to save the crop. Correct hilling, along with a liquid solution to burn or kill the weeds, can be very effective. The liquid does not necessarily have to be an herbicide for controlling weeds in corn. An application of five to ten gallons of liquid 28-0-0 nitrogen

Fig. 91A

Fig. 92

directed into the row will burn most weeds. The nitrogen will also burn corn leaves, but this simply gets rid of a few bottom leaves which in no way hurts the yield. The burning, along with the covering action of hilling, can very effectively control weeds. As the weeds decompose they release their nutrients and serve in this respect as a green manure crop. The extra nitrogen, strategically placed in conjunction with the many benefits of decay, can be beneficial to the corn crop also.

Weeds in corn rows may also be burned back with three gallons of concentrated sulfuric acid per acre in thirty gallons of water directed to the base of the corn plant. This will be outlined at the end of this unit.

Rolling cultivator gangs, or disc hillers, are effective for hilling. It is an advantage if the hiller units can be adjusted to either directly oppose each other or if one can be set to lead the other. By staggering the hillers, the front or leading unit lays the weeds over and the trailing unit completely covers them. For normal cultivation the units could be readjusted to directly oppose each other. Hilling is an asset and its capability should be built into the cultivator unit.

NUTRIENT APPLICATION

Finally, a cultivator should have the capability of applying a liquid or dry nutrient close to the plant while cultivating. In effect, this is a sidedressing operation. The key, however, is to place the nutrient close to the plant and not in the center of the row. Liquids have some advantages over dry, but either can be used.

This third function alone can make cultivation a profitable operation and can be the link that completes the chain in a most economical soil fertility "system." My field staff and I have pioneered and fine-tuned this method of sidedressing and consider it a very profitable way to get maximum mileage for fertility dollars spent. This system will be fully explained in the unit on "Soil Fertility."

In summary, we must conclude that cultivation has more benefits than just the control of weeds. It may very well be the single most important tillage operation you can perform. It could be the best salvage tool at your disposal because it can salvage the crop from both weeds and poor soil aeration while placing nutrients very strategically and with the requisite timeliness for maximum economy and effect. We could want nothing more from a tool.

As we close this unit we must mention how detrimental improper cultivation can be. Too often farmers will set their shovels too deep and too close to the row. The corn plant can actually be seen rocking back and forth as a result of the shearing off of a solid wall of roots on both sides. We simply cannot afford to severely prune the roots on an annual crop.

In Figure 91 we see a rather poor approach to cultivation. This type of tine does a good job of working between the rows, but the tines are working too close and too deep next to the row. This cultivator is not equipped to hill, it cannot go deep between the rows and it is not equipped to apply a nutrient. It may require the re-engineering of your present cultivator, but do not neglect the three important functions of cultivation, deep-tilling between the rows, hilling and nutrient application.

SULFURIC ACID FOR WEED CONTROL IN CORN

We discussed briefly the use of non-herbicides such as liquid nitrogen to burn back weeds that may be growing out of control in the corn rows. At this time we are also experimenting with the use of sulfuric acid as a means of controlling weeds within the corn row.

For this purpose we are using two gallons of concentrated sulfuric acid per acre in thirty to thirty-five gallons of water and sprayed in a band at the base of the corn plant. We are including a sticker-spreader agent which makes the acid adhere to the surface of weeds such as velvet leaf. To date we are excited with our results and trying it on select farms that either prefer not to use herbicides or simply want to cut weed control costs.

At the present time the bulk costs for sulfuric acid run from $1.30 to $1.60 per gallon. The sulfuric acid contains about thirty-two percent sulfur in the plant-usable, sulphate form. A gallon of sulfuric acid contains nearly one dollar's worth of sulfur, making it a very low-cost product. We think the acidity effect (on the soil) of two gallons of sulfuric acid per acre is of little concern.

Sulfuric acid is dangerous and difficult to handle. It is damaging to skin, clothes and equipment. For its application one should take the safety precautions of wearing gloves, goggles and keeping a supply of water close by to rinse or neutralize spills, especially on the body. Fiberglass tanks, fiberglass or acid-resistant pump housing and impellars, nylon spray tips and acid-resistant rubber on plastic hoses must be used.

A very pungent sulphide odor will be noticeable in the fields for a day or two. We believe it is possible this may even repel insects and are experimenting with this idea also.

The acid also burns the lower corn leaves, but this is no problem. In fact, corn often produces better if some of these lower leaves are removed as they contribute little to production. But, should the acid get inside the corn whorls, it would be fatal to the plant. We consider the use of sulfuric acid on plants that can afford to lose their bottom leaves such as corn. The acid would kill soybeans.

We are not recommending a mass switch to sulfuric acid, but use on a limited basis as an alternative means of salvaging a problem situation. One should tread lightly in using new ideas such as this by testing on limited acreage and increasing with experience.

END ROW MANAGEMENT

As a final footnote on tillage we must mention that the end rows are probably the most physically neglected areas on the entire farm. By turning around on them several times during the season and using them as roadways during harvest we impart a tremendous amount of compaction. As a general rule end rows are treated as a necessary evil and little thought given as to their problems and needs.

Instead we must view end rows as important areas of production and recognize that compaction in the form of wheel tracks is much more severe on these areas than anywhere else in the field. Soil compaction is generally the number one problem within our fields and is especially bad on the end rows. Soil compaction deprives the soil of an essential ingredient, air. Soil life, root growth and nutrient recovery are oxygen-demanding processes and their function relates directly to the soil's ability to breathe.

Compaction also impedes water management which further compounds soil-plant problems. Runoff from compacted end rows can often over-saturate the soil for several feet within the field to broaden the problem area. On end rows we do our poorest management of the two most essential soil basics — air and water.

Because the primary problem is wheel-track compaction we believe that deep chiseling is the simplest and most economical way to alleviate the condition. By eliminating the plow and properly removing the pans, the inner fields may not require additional subsoiling for years. The end rows, however, should be subsoiled annually as the final tillage trip after harvest. They should be deep-tilled again between the rows immediately after planting or as the final operation after cultivation, spraying or sidedressing.

The chiseling depth can be decided by soil type and conditions during seedbed preparation and planting. Wheel tracks impart deeper compaction on heavy-textured soils carrying optimum moisture. Normally a chiseling depth of fourteen to sixteen inches in the fall and ten to twelve inches during the final spring operation will do a good job on most soils.

The deep chiseling should be done with a relatively narrow tine as outlined in the unit on "Water Management." The principles are the same; our goal is to slot the soil to permit water penetration and the lateral diffusion of soil gasses.

Another consideration is the use of some of the better surfactants presently on the market. Some of these products improve water movement and may have a use on problem areas such as end rows. End rows may provide a good testing site to determine their worth. Generally we do not recommend bathing the entire field with surfactants because we simply cannot predict their long-term effect on certain nutrients, etc. Their best areas of use may be on severe problem areas or in specialized placement.

A good farm production program defines and solves problem areas. Problem solving is the key to success in any business. Since end rows are problem areas and may represent considerable acreage, it would appear that an "end row" management program would be good business.

BASIC NUMBER FOUR — MANAGING SOIL FERTILITY

We stand back and proudly view a corn field as a thing of beauty — the pinnacle of perfection. But with an apparent lack of any deficiency signs, why did it produce one-hundred thirty bushels of corn instead of two-hundred fifty? We must ask ourselves, is American agriculture efficient? The answer is "yes" in terms of production per man unit, but probably a resounding "no" in terms of our actual yield potential.

Soil fertility is synonymous with the term "food" because both imply the containment of nutrients. The difference is that we usually relate fertility to plants and food to animals.

Nutrient presence or containment, however, does not guarantee usage by either the plant or animal. Food must be both ingested and digested before the individual nutrients can become available and absorbed into the blood stream of an animal.

The process for plant nutrition is very similar. The soil can be considered the plant's stomach. Digestion must take place within the soil system before the nutrients within the complex organic matter and fertilizers can be converted into plant-usable forms. After these nutrients are changed to forms available to the plant they must be absorbed into the plant system and converted into meaningful production. The recovery of soil nutrients by the plant is the key to the expression of soil fertility.

We learned earlier in the text of the role of soil oxygen in root growth and nutrient uptake by the roots. Both are oxygen-demanding processes. This is why soil aeration is a major factor in determining recovery of nutrients by the roots.

Soil water management is also important. In the study of soil fertility we will emphasize again and again that the fertility of a soil must ultimately be judged on the ability of the plant to obtain its nutrients.

Management of soil fertility is more complicated than simply measuring and applying nutrients. In crop production the presence of a soil nutrient is important, but its recovery is essential.

Soil fertility is not simply a measure of soil levels of the individual nutrients, but instead is a function of plant response. The whole purpose of managing and manipulating soil chemistry is to induce a favorable plant response. A favorable plant response may be defined as a proper return on our fertility investment. This involves the intelligent **management** of both **application** and **recovery.**

Management of soil chemistry includes four important areas of expertise. These are:

1. Use of soil testing technology to determine the available or exchangeable soil nutrient levels;
2. Interpretation of the soil's chemical problems from the laboratory data;
3. Correction of the soil's chemical needs;
4. Application of cultural practices that will assure recovery of the soil nutrients by the plant.

Soil testing is a necessary first step in the management of soil fertility. There are three important areas to consider in securing meaningful soil analysis data.

First, we must gather representative soil specimens during the soil sampling operation. Second, we must select a laboratory of excellent quality, and third, we must run a complete soil analysis or audit. Correct interpretation is not possible unless accurate soil analysis data is available.

SOIL SAMPLING

The person who is doing the sampling must be sufficiently schooled and not simply sent to the field to bring back a little dirt. In fact, the farmer should take the time to do his own sampling or else accompany the person designated to do it. Who is better qualified or knows the farm better than the farmer himself? It gives him a chance to get down from his equipment to study his greatest asset in more detail. He can kick the soil, feel it, smell it, look for worms, insects, undecayed residue, soil colors, hard pans, plow pans, density of wheel tracks, stalk quality, etc. It should be made an investigative experience. Too often all that is seen of the soil is its surface from the cabs of tractors.

Soil sampling requires a minimum investment in equipment. A soil probe is needed, soil sample bags, a clipboard and a farm map, plus a bucket or pail in which to carry everything. A five-gallon pail works very well and a soil probe is more accurate and easier to use than a shovel or spade.

Figure 92B shows how to sample a field. This particular field has three major soil areas which must be sampled separately. The number of acres per soil sample is not too important. A single composite soil sample may represent twenty or more acres. Small, individual fields or problem areas may be sampled separately, and in such circumstances a sample could cover a relatively small area. An important thing to remember is to sample each major area separately.

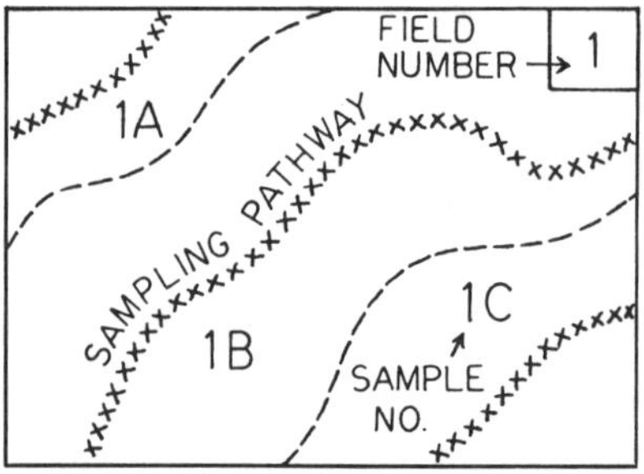

Fig. 92B

Figure 92B also shows a sampling pattern that should give an average field fertility value. It shows how the field should be numbered. Sample areas and the pathway used to gather them are marked. The same path should be followed in future sampling.

In Figure 93 we see how to correctly label the soil sample bag. This one belongs to R. Jones and is Sample A from Field Number 1. Labeling should be kept as simple as possible.

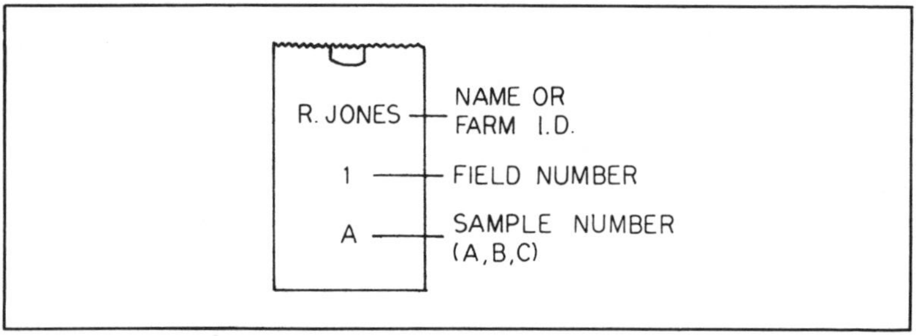

Fig. 93

DO'S AND DON'TS OF SAMPLING

Take fifteen to twenty probings, seven to eight inches deep, from each sample area. Place soil directly into the pail or bucket, mix it well and put a representative sample into a soil bag. Do not grind or dry the soil. A good laboratory is equipped to do this. Depth is important, as we will see when we get into fertility recommendations.

Correctly label the map and sample bags. Do this immediately to prevent errors. Always print because too often the only person who can read one's field writing is himself.

Do not probe in starter fertilizer bands or directly into wheel tracks. Probe eight to ten inches from the row and avoid the wheel tracks if possible.

If fertilizer has been spread ahead of sampling remove one-half to one inch of soil before probing. If probing into sod, simply remove the residue from the top of the probe before putting the soil into the bag.

Keep each sample as uniform as possible. Soil from clay and sandy areas should not be mixed.

Avoid unusual areas. Do not include a waterway or ditch bank in a regular sample. Such areas should be sampled separately. Always try to sample the average of a given area. The basic guidelines of sampling are simply common sense and conscientiousness, and, for these reasons you may want to do your own.

SELECTION OF A SOIL TESTING LABORATORY

The accuracy of lab work is determined by the equipment, techniques, personnel and quality control. Equipment for soil testing can vary in cost from fifty-dollar "do it yourself" home-testing kits to equipment costing many thousands of dollars. Because accuracy and consistency will vary you must select a competent laboratory if you are to trust the test data results.

There are several good laboratories located throughout the USA. The many commercial consulting firms, such as Advanced Agriculture, Inc., depend very much on the quality of their lab data. By sending thousands of samples, retesting farms year after year and sending duplicate samples for comparison, a firm can constantly monitor the accuracy of a given laboratory.

In this study of soil test interpretation and fertility recommendations we will use the data from A and L Great Lakes Agricultural Laboratories, Inc. Because Advanced Agriculture, Inc. uses this lab the author is most familiar with their standards, procedures and format. A thorough understanding of the lab data is important for making sound and accurate recommendations. A and L Great Lakes Agricultural Laboratories, Inc. is recommended as one of several outstanding analytical laboratories.

From the lab data a soil consultant should be able to list your problem areas in their relative order of importance. If a consultant does not list your problems, or attempts to make recommendations from partial data, you should question his qualifications and/or abilities. You should not be expected to accept anything less because your business and livelihood are at stake. Your fertility decisions must be based on knowledgeable recommendations from soil data that is as comprehensive as it is possible to obtain. You will get no more than you demand and your knowledge will determine your level of acceptance.

Because it is difficult to make demands in areas we do not fully understand, our study of soil chemistry will shed light on this area of confusion. In fact, we believe that you, the farmer, must be able to make your own recommendations from your soil test data. You must become the expert. The time when you could tell your fertilizer dealer to, "put on enough fertilizer for a 150 bushel corn yield," is gone.

You should no longer be victimized by anyone selling an expensive "corn fertilizer" or by paying premium prices for fertilizers with a sprinkling of trace elements. There is no such thing as a "corn fertilizer" that fits all soil types. Furthermore, you should not pay a premium for low levels of three or four basic trace elements when the problem in your particular soil might be low zinc. If zinc is low you must be aware of it and invest in solving the problem.

Such sales gimmicks are an affront to our intelligence. How much has been spent through the years for such gimmicks? One of the purposes of this unit on soil fertility will be to gain enough expertise so that our fertility inputs will be based on real needs.

Having properly sampled, selected a good laboratory, and run a complete test, we are now ready to interpret the data, define the problems and take corrective action.

UNDERSTANDING AND INTERPRETING THE SOIL TEST DATA

The soil is many things, but most important, it is a storehouse of plant nutrients. In this respect we consider it a pantry. The soil test is simply a means of chemically walking into the pantry and taking an inventory.

These inventory figures can be compared to established standards to define soil nutrient problem areas. The pantry must be filled with a balance of nutrients. We would not want our kitchen pantry filled with potatoes and flour alone and it is the same with the soil pantry. The purpose of soil testing is to determine needs and to ultimately bring the soil pantry into a reasonable balance of soil nutrients.

Before entering an in-depth study of soil chemistry we want to caution against any mental blocks the subject may evoke. You do not need a degree in soil chemistry to learn the basic soil chemistry necessary for interpretation of soil test data. All that is needed is a logical, step-by-step presentation of the subject and a desire to become an expert, so let us begin.

In our study we will use the A and L Agricultural Laboratory's Soil Analysis Report Form. We will study it item by item in a sequential pattern that will make it easy to follow and understand.

All laboratory data must be compared to certain generally accepted standards. In this respect you will note that most laboratories assign a rating to their data. Let us take a moment to think about and analyse these rating systems. What is a high reading for a given nutrient? Is it possible for a high soil test reading to actually turn out to be low? This is certainly possible.

Remember, we judge a soil's fertility not only by soil nutrient levels, but also by the plant's ability to recover the nutrients. We cannot accept the lab ratings as "Gospel" because there are several ways that a plant can become deficient in respect to a given nutrient even though the soil test may show high levels for that nutrient.

Figure 94 shows three separate root systems of a crop. It shows small, medium, and large root systems. A high soil test level may actually be high for the large root system, but could turn out to be low for the small root mass.

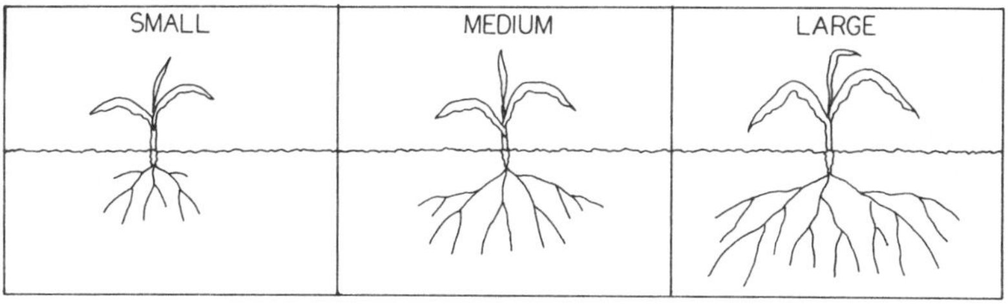

Fig. 94

Another example of a soil situation that could affect the rating system is shown in Figure 95. Soil testing and recommendations are based on a depth of sampling which is usually seven and one-half to eight inches. It is estimated that about ninety percent of the feeder roots of most crops are found in this soil depth. This illustration shows three inches of the seven and one-half being comprised mainly of clods and no roots. Because of this, three-sevenths of the tested and

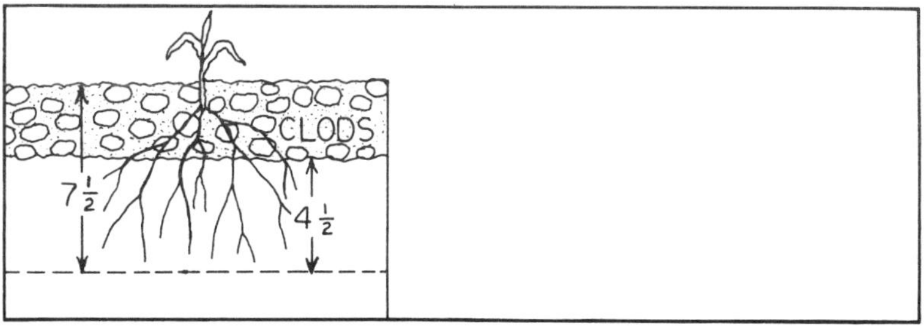

Fig. 95

fertilized zone is out of action and the high rating may turn into a low rating with regard to crop production.

The lab rating is necessary and we are not finding fault with the system, but simply emphasizing that soil fertility relates to crop response as well as to soil levels. We must continually remind ourselves that when we apply soil nutrients our task has just begun. The ability to manage soil air, water and residue will ultimately determine the crop's response to soil fertility levels.

Most laboratories offer computerized soil fertility guidelines based on the soil test results. As you study this unit on soil fertility you will recognize that computers have no imagination. Fertility recommendations demand an in-depth knowledge of the individual farm situation. Budgets, drainage, equipment, water-holding capacity, timing, tillage, residue management, manure systems and placement of fertility are some of the factors that must be considered in a sound fertility management system. No computer is programmed to this degree. Computer programs can be guidelines only and must be adjusted to fit each individual situation.

Figure 96 is an A and L Agricultural Laboratory's Soil Analysis Report Form. The top half of the form contains the items referred to as the "basic soil test." In the lower half we are mainly concerned with sulfur and the five trace elements.

Each soil sample is normally tested for all of the items in the top half except sodium. Sodium is usually tested only if the sample has come from an area of the country which is high in sodium. If soils were very alkaline we would instruct the lab to run the sodium bicarbonate test for phosphorous instead of the P1 and P2 normally run. Note the sodium bicarbonate formula ($NaHCO_3$-P) in the P2 column.

In the lower half of the form are the test columns for sulfur, zinc, manganese, iron, copper and boron. In general, the nitrate test, excessive lime, soluble salts, chloride, molybdenum and particle size analysis are special tests that are not normally run for general agriculture. Take a moment at this point to familiarize yourself with the different columns on the soil analysis form in Figure 96.

Figure 97 shows the A and L Laboratory's Soil Sample Information Sheet. This form provides the laboratory with the necessary information to run the soil analysis and it should accompany the soil samples sent to the lab. Most labs have similar information forms that they make available upon request.

You will note that the **Standard 1A Column** is marked for all samples. This will provide the tests on the top half of the soil analysis sheet as shown in Figure 96. This is the standard soil test that should be run on all samples.

REPORT NUMBER

A & L GREAT LAKES AGRICULTURAL LABORATORIES, INC.
5011 Decatur Road • Fort Wayne, Indiana 46806 • Phone: (219) 456-3545

SEND TO: GROWER: SAMPLES SUBMITTED BY:

DATE OF REPORT: PAGE:

SOIL ANALYSIS

SAMPLE NUMBER	LAB NUMBER	ORGANIC MATTER		PHOSPHORUS		POTASSIUM	MAGNESIUM	CALCIUM	SODIUM	pH		Cation Exchange Capacity	PERCENT BASE SATURATION				
		%	ENR lbs/A	BRAY P1 ppm-P	BRAY P2 ppm-P	K ppm	Mg ppm	Ca ppm	Na ppm	SOIL pH	BUFFER pH	meq/100g	% K	% Mg	% Ca	% H	% Na

EXAMPLE

SAMPLE NUMBER	SULFUR S ppm	ZINC Zn ppm	MANGANESE Mn ppm	IRON Fe ppm	COPPER Cu ppm	BORON B ppm	SOLUBLE SALTS mmhos/cm				COMMENTS

BY

CODE TO RATING VERY LOW (VL), LOW (L), MEDIUM (M), HIGH (H), VERY HIGH (VH) ENR ESTIMATED NITROGEN RELEASE

Our reports and letters are for the exclusive and confidential use of our clients and may not be reproduced in whole or in part, nor may any reference be made to the work, the results, or the company in any advertising, news release or other public announcements without prior written permission.

Fig. 96

If the **Micronutrient Column 3** is marked, the lab will run sulfur and the five trace element tests shown on the bottom half of the soil analysis sheet. There are other test columns and individual tests that are available and you should have your laboratory mail to you a copy of their "**Schedule of Fees.**" From this schedule you can determine your tests and their cost.

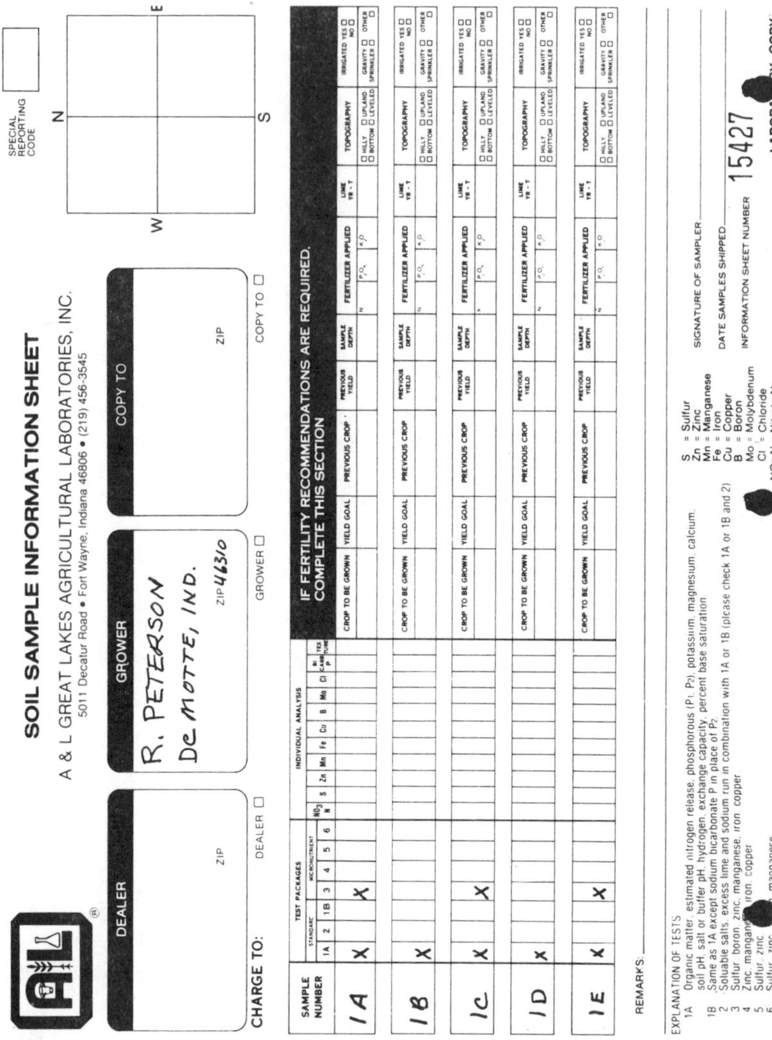

SOIL SAMPLE INFORMATION SHEET

A & L GREAT LAKES AGRICULTURAL LABORATORIES, INC.
5011 Decatur Road • Fort Wayne, Indiana 46806 • (219) 456-3545

SPECIAL REPORTING CODE

DEALER — ZIP

GROWER — R. PETERSON, DeMOTTE, IND. ZIP 46310

COPY TO — ZIP

CHARGE TO: DEALER ☐ GROWER ☐ COPY TO ☐

SAMPLE NUMBER	1A	2	1B	3	4	5	6
1A	X			X			
1B	X						
1C	X			X			
1D	X						
1E	X			X			

IF FERTILITY RECOMMENDATIONS ARE REQUIRED, COMPLETE THIS SECTION

CROP TO BE GROWN | YIELD GOAL | PREVIOUS CROP | PREVIOUS YIELD | SAMPLE DEPTH | FERTILIZER APPLIED | LIME YR-T | TOPOGRAPHY | IRRIGATED

REMARKS:

EXPLANATION OF TESTS

1A Organic matter, estimated nitrogen release, phosphorous (P_1, P_2), potassium, magnesium, calcium, soil pH, salt or buffer pH, hydrogen, exchange capacity, percent base saturation

1B Same as 1A except sodium bicarbonate P in place of P_2

2 Soluable salts, excess lime and sodium run in combination with 1A or 1B (please check 1A or 1B and 2)

3 Sulfur, boron, zinc, manganese, iron, copper

4 Zinc, manganese, iron, copper

5 Sulfur, zinc

6 Sulfur, zinc, ... manganese

S = Sulfur, Zn = Zinc, Mn = Manganese, Fe = Iron, Cu = Copper, B = Boron, Mo = Molybdenum, Cl = Chloride, NO3-N = Nitrate N

SIGNATURE OF SAMPLER

DATE SAMPLES SHIPPED

INFORMATION SHEET NUMBER 15427

LABORATORY COPY

Fig. 97

ORGANIC MATTER AND ENR

An area of importance on the report is the Organic Matter Column. This shows two figures and the first figure is the percent of humus with the other indicating the Estimated Nitrogen Release (ENR) in pounds per acre. These two tests are automatically run on all samples.

The terms, "organic matter" and "humus" should not cause confusion. Keep in mind that humus is organic matter, but organic matter is not always humus. Humus represents the end product that remains after organic matter has been decomposed. The Organic Matter Column shows the percent of **humus** in your soil.

Humus, which gives soil its black color, is a very important constituent. A major goal in soil management is to stabilize humus and gradually increase it instead of slowly depleting it. Humus has many roles. It binds soil particles and increases soil aeration and water-holding capacity. It provides a home for soil life and is the fraction of soil with the most fertility-holding capacity. In this respect humus is a large pantry.

The amount of humus produced in the decay process is determined by the type of organic material with which it began. The more complex the carbohydrate is, the higher the humus yield will be. Lignin and cellulose materials produce high humus yields. Green, lush plants consist mostly of water, protein and sugars. They contain very little, if any, lignin and cellulose and result in very little humus after decaying. Highly lignaceous materials such as cornstalks and small grain straws are the humus producers.

It would require a large book to cover the chemistry and importance of humus. It suffices to say that we must recognize its importance and make every effort to implement humus building practices.

The Estimated Nitrogen Release (ENR) figures are determined with the assumption that each percent of humus will release approximately thirty pounds of nitrogen per acre each year. But, there are apparent differences in the age of humus, its degree of oxidation, the original matter comprising the humus, etc. For these reasons these figures are not too dependable.

We have seen ENR figures as high as two-hundred fifty pounds of N per acre and still seen corn fail if additional nitrogen was not applied. Certain muck soils can grow corn without additional nitrogen, yet it will fail on other seemingly similar muck soils. In summary, these figures may hold true on virgin soils, but possibly not on soils that have been farmed for many years. Testing of nitrogen to determine your most economical rates will be discussed in detail in the unit on **Nitrogen Management.**

PARTS PER MILLION (PPM)

On the Soil Analysis Report nearly all figures are given in Parts Per Million (PPM). PPM must be understood because soil fertility needs and rates of application are based upon it.

When sampling soils for general farm crops we try to probe to a depth of seven and one-half to eight inches. On the average, a soil layer seven and one-half inches deep, over one acre, will represent two million pounds. We must know this figure to fully comprehend that 1 PPM represents two pounds per acre. In other words, we must add two pounds to two million pounds of soil to increase the soil level by 1 PPM per acre.

You will note that the soil analysis figures are given as elemental PPM, such as calcium (Ca) and boron (B), etc. To increase soil boron levels by 1 PPM we have to apply two pounds of elemental boron per acre. All we must remember is that it requires two pounds of any element to increase the soil level by 1 PPM. Needless to say, the two pounds added cannot be tucked away in one spot, but instead must be thoroughly mixed throughout the two million pounds of soil before it will accurately reflect on a soil test. Also, if we till deeper than seven and one-half inches we must increase the application rates accordingly.

A crop can respond to a small trace element application even though it may not show up immediately on a soil test. The soil levels can be raised by placement of material in the starter bands. The crop will respond, but the application would not show up on a soil test taken away from the band. This must be understood when we work with small applications of trace elements and will be explained more thoroughly in the section on **Trace Element Management.**

CATION EXCHANGES CAPACITY AND BASE SATURATION

The Cation Exchange Capacity will hereafter be called the "CEC." The meaning of CEC must be understood as it relates to the knowledge of basic soils, fertility application and fertility placement systems. The Percent Base Saturation columns to the right of the CEC Column have a very important relationship to the CEC.

Let us think of the CEC as a measurement of a soil's pantry size or fertility-holding capacity, for one acre that is seven and one-half inches in depth (two million pounds of soil).

On a soil test the CEC of a sample is assigned a number such as 5, 2, 10, 4, etc. The number represents the pantry or room size which will hold fertility, with the larger numbers representing the larger pantry sizes.

In Figure 99 note that a 10 CEC soil has twice the fertility-holding capacity or pantry size of a soil with a 5 CEC.

When viewing the CEC as a pantry or room we realize that a room can be small, large, or any size in between. Some soils have a huge

5-CEC	10-CEC

Fig. 99

pantry comparable to a gymnasium while others are as small as a linen closet.

Why is it called **cation exchange capacity?** It is now time to review a few chemical terms. In chemistry elements or molecules are called either **cations** or **anions.** A **cation** is a positive (+) charged element while an **anion** is a negative (–) charged molecule. Opposite charges (+ and –) attract each other, while like charges such as + and +, or – and –, repel each other. Plants use both cations and anions. These are listed in Figure 100.

CATIONS (+) POS.	ANIONS (–) NEG.
Ca^{++} CALCIUM	NO_3^{-} NITRATE
Mg^{++} MAGNESIUM	$PO_4^{\equiv}$ ORTHO PHOS
K^{+} POTASSIUM	$SO_4^{=}$ SULPHATE
NH_4^{+} AMMONIUM	
H^{+} HYDROGEN	
Na^{+} SODIUM	
Cu^{++} COPPER	
ZN^{++} ZINC	
Mm^{++} MANGANESE	

Fig. 100

The cation exchange capacity (CEC) implies that only cations, or positive-charged elements, are in the soil pantry. If only positive-charged elements (cations) stick to the pantry the pantry itself must be negative-charged. In other words, soil particles carry negative sites that attract positive-charged elements (cations). We often refer to the soil as "ground," possibly because it is negative-charged.

In effect the CEC is a measurement of the number of negative sites in one acre of soil, again considered to be a depth of seven and one-half inches, or two million pounds of soil. A greater number of negative sites means a larger CEC number, or simply stated, a larger pantry capacity.

Figure 101 shows a diagram of how we may imagine a soil particle to look. Notice the many negative sites. Each negative site will attract positive-charged elements until all of the sites are filled. There will be no bare negative sites. The cations that fill the sites will determine the soil's chemical balance. Liming and fertility practices will

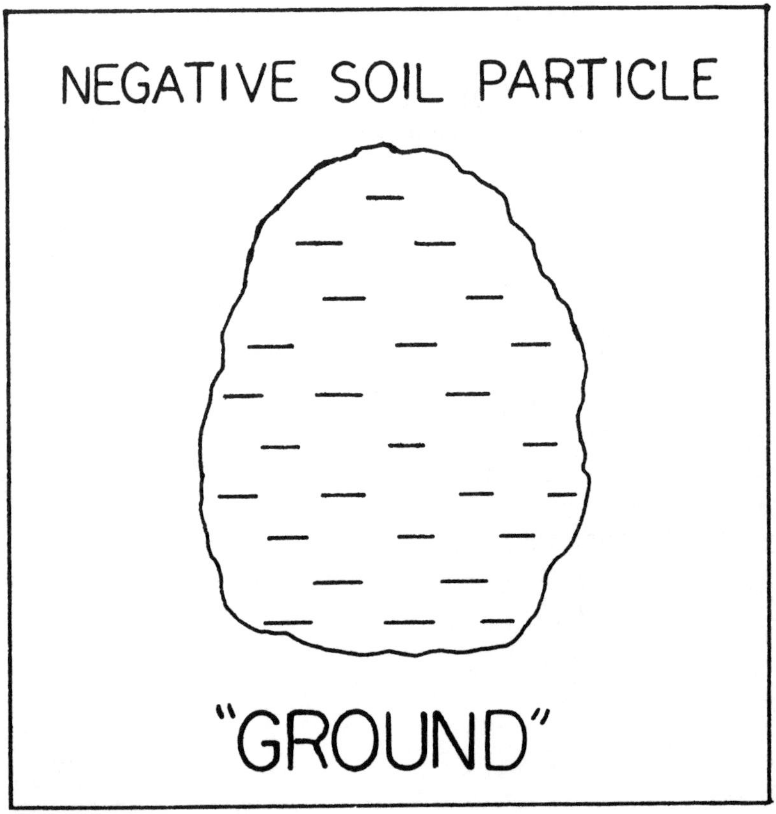

Fig. 101

determine what elements occupy these sites. **Again, we must emphasize that the pantry will be completely filled with cations of one kind or another.**

Because the CEC numbers vary considerably from one soil type to another the soil makeup will also vary. Soils contain sand, silt, clay and humus, either as single ingredients or in various combinations. All of these ingredients contain or carry various levels of negative sites and either singly or in combination make up the soil's pantry or CEC.

As single ingredients, sand, silt, clay and humus each have their own pantry size, or CEC number, which varies greatly, as shown in Figure 102. We can readily see that sand and silt have very small pantries while humus would resemble a gymnasium in comparison. Clay varies because it may have either two or three-layered particles. Nevertheless, clay has a large fertility-holding capacity. The individual soil sample CEC numbers are determined by the humus and soil particle combinations.

C.E.C. RATING GUIDE FOR THE VARIOUS SOIL FACTIONS	
SAND	5 TO 2 C.E.C.
SILT	2 TO 5 C.E.C.
CLAY	UP TO 60 C.E.C.
HUMUS	UP TO 250 C.E.C.

Fig. 102

It is of interest to note that the only practical way to increase the CEC, or pantry size, would be to increase humus levels. Clay could be incorporated on a sand hill, but that would not be very practical! The increase of humus gives many other benefits as well.

Pantry size does not necessarily relate to the ability of a soil to produce. Air and water, as noted, always remain critical to production regardless of pantry size. For instance, a large pantry of pure clay is of no use if air and water cannot be managed. Do not worry too much about your CEC number. In the short run there is little that can be done to alter it. Be more concerned with managing air, water, and slowly building humus and nutrient levels, along with other practices that will be supportive of the soil.

Part of the soil test concerns taking an inventory of the cations filling the pantry. Figure 103 shows how a soil particle holds cations.

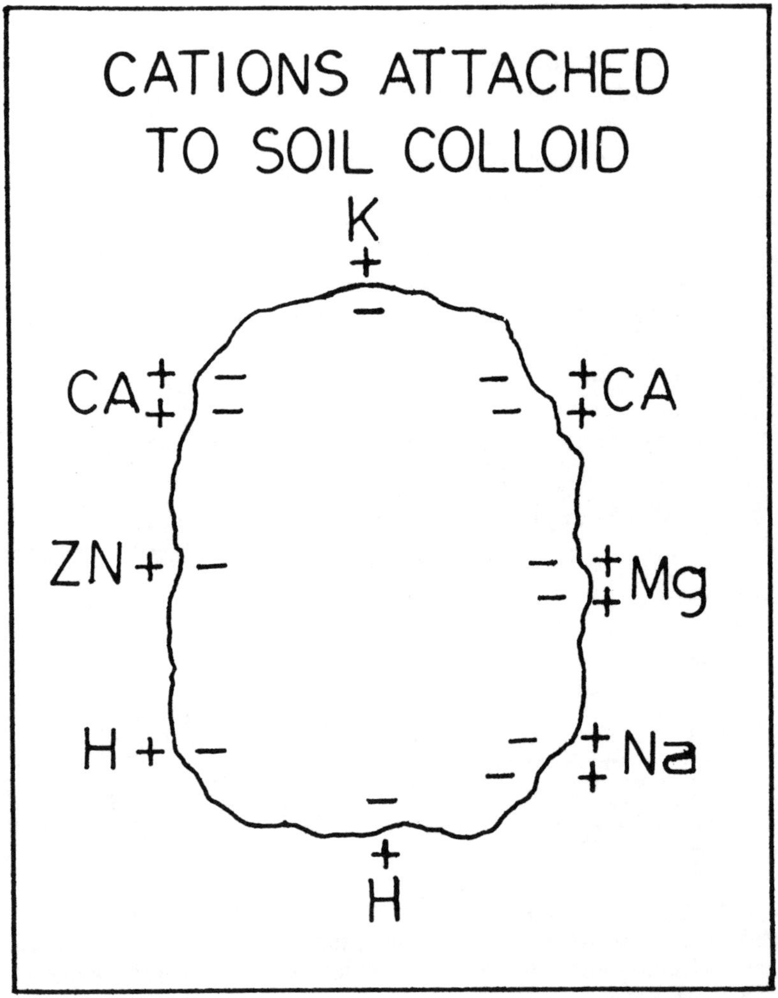

Fig. 103

In addition to trace elements the cations found in the pantry are potassium (K+), magnesium (Mg++), calcium (Ca++), hydrogen (H+) and sodium (Na+). The soil test tells how much of these elements are in the soil and also what percent of the pantry is occupied by each cation.

The **Percent Base Saturation** figures on the soil test tell how the pantry is filled. No one knows for certain what a perfect pantry balance would be, but Figure 104 shows what is normally considered to be a good "ballpark" balance.

Trace elements are in such minute quantities that they are measured in PPM instead of a percent of pantry fill. Sodium (Na) is very low in most soils and usually not even tested unless it is present in excessive amounts. However, in some soils the sodium base saturation can actually exceed potassium saturation.

You will immediately notice in Figure 104 that calcium dominates the pantry. Its percentage range also varies the most. This will be explained later.

The Base Saturation Figures on a soil test are very important because any cation imbalances can be seen at a glance. The negative sites in the soil are totally filled. All of the soil cations have either one or two positive charges. A calcium molecule (Ca^{++}) has two positive charges so it will occupy two negative sites on the soil colloid.

By knowing the cation exchange capacity of our soils we can calculate the number of pounds of any cation needed to fill a certain percentage of the pantry. It requires different amounts of each cation to completely fill a single CEC.

In Figure 105 we show the pounds of each cation required to completely fill one CEC. If any one CEC contained four-hundred pounds of calcium (200 PPM) per acre, there would be no room for anything else.

As an example: How many pounds of calcium would be needed per acre to have a seventy percent base saturation on a 10 CEC soil? Each single CEC would require four-hundred pounds (200 PPM) of calcium per acre, as shown in Figure 105, to completely fill the pantry. So, it would require four-hundred pounds of calcium multiplied by 10 CEC, or a total of four-thousand pounds of calcium to fill the pantry one-hundred percent with calcium. It would require twenty-eight hundred pounds of calcium (1400 PPM) to fill the pantry to the desired seventy percent fill ($4000 \times 70\% = 2800$ pounds).

An important thing to remember is that the soil test shows calcium in PPM. A 10 CEC soil would require fourteen-hundred PPM on the soil test to equal twenty-eight hundred pounds per acre, for a base saturation of seventy percent. One PPM on the soil test is equal to two pounds per acre because we are working with two million pounds of soil.

While calcium has been presented as the example, magnesium and potassium would be calculated in the same way. We need not be concerned with the calculation of hydrogen or sodium.

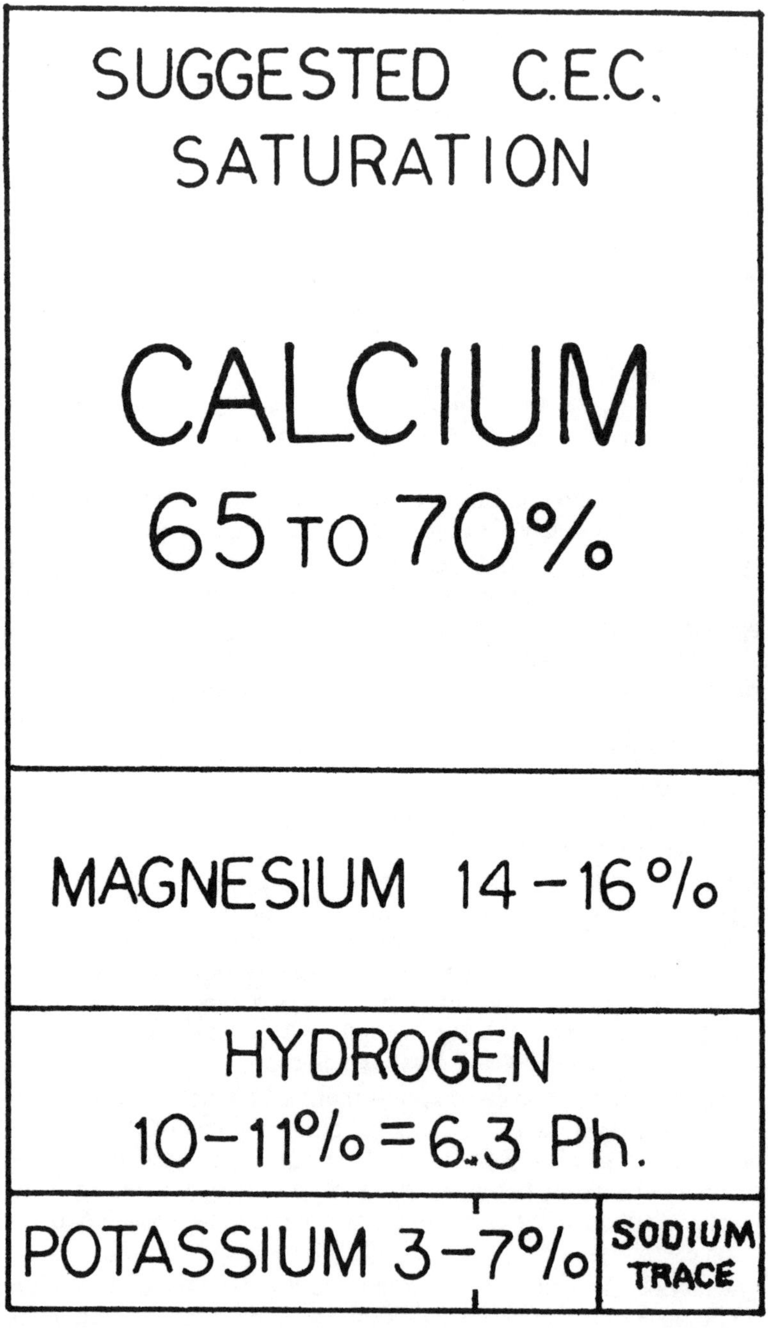
SUGGESTED C.E.C.
SATURATION
CALCIUM
65 TO 70%
MAGNESIUM 14-16%
HYDROGEN
10-11% = 6.3 Ph.
POTASSIUM 3-7%
SODIUM
TRACE

Fig. 104

CATION	LBS. TO FILL 1 CEC
CALCIUM	400 LBS.
MAGNESIUM	240 LBS.
POTASSIUM	790 LBS.
AMMONIUM	360 LBS.
HYDROGEN	20 LBS.

Fig. 105

By calculating the pounds per acre needed for a desired pantry fill and comparing this to your actual soil test levels you will know whether a given cation is deficient or excessive.

In summary, there are several points that will help clarify cation exchange capacity (CEC):

CEC balance is important. The pantry will hold only a specified amount and overfilling with one cation is done at the expense of another. Deficiencies are always easier to manage than excesses. This is another reason why applications of fertilizers must be based on soil test data.

On some soils it is possible to have a very low CEC which cannot hold enough nutrients to grow a high yield. Nutrients will leach readily if applied beyond the soil's ability to hold them. This factor also influences fertility application methods and amounts.

The amount of nutrient required per acre to raise the percent base saturation increases as the CEC increases. In other words, it will require more pounds of a nutrient such as potassium to fill the pantry to a three percent level on a 10 CEC soil than it would on a 5 CEC soil.

While we must understand the cation side of plant nutrition and cation exchange capacity, it is also necessary to know about the anion side of plant nutrition. Anions, as you will recall, are negative-charged molecules and do not attach to the negative-charged soil particles.

In plant nutrition we are concerned with the anions sulphate ($SO_4=$), phosphate ($PO_4\equiv$) and nitrate ($NO_3=$). All of these carry negative charges and are in the oxide form instead of being in the elemental forms such as the cations calcium, magnesium and potassium.

Sulphate and nitrate leach easily from the soil. To protect against leaching these anions can combine with certain elements that will greatly reduce their leachability.

Plants cannot utilize the elemental forms of sulfur, phosphorous and nitrogen. However, they do readily utilize the anion forms as sulphate (SO_4=), ortho-phosphate (PO_4≡) and nitrate (NO_3=). It requires soil microbiology to convert sulfur, phosphorous and nitrogen to their oxide, or plant-usable forms.

With the preceding information we are now ready to study the soil test data item by item.

PHOSPHOROUS

Information on phosphorous can be found on the soil test in the two columns P1 and P2. The P2 will always be a larger number than the P1. In the testing process the lab uses a stronger phosphorous extraction acid for the P2 test making it more nearly a measurement of the total phosphorous in the soil. It measures the water-soluble, weak acid soluble (P1) and some of the less available phosphate reserves.

The P1 is extracted with a weaker acid and shows only the readily available or root soluble soil phosphorous. The P1 and P2 readings are given in PPM of elemental phosphorous (P). For general crops we believe the P2 level should be 50 PPM or higher with a P1 level of 25 PPM or more. A soil test reading showing the P1 level at fifty percent of the P2 would indicate very good phosphorous availability. A soil with a P2 readout of 50 PPM and a P1 of 25 PPM would have good base phosphorous levels for general crops and we would be concerned only with its maintenance.

Under the P2 Column on the soil analysis form you will note the formula $NaHCO_3$-P. This is the formula for the sodium bicarbonate phosphate test. If the soil pH exceeds 7.0 the **sodium bicarbonate** test should be run on the soil samples instead of the P1 and P2 tests. Instead of checking the BASIC TEST S1A on the laboratory information sheet accompanying the samples to the lab, the BASIC TEST S1B would be checked. The laboratory analysis figures and ratings are quite different for the sodium bicarbonate test. It is a more accurate test under alkaline conditions.

All phosphate fertilizers, when added to the soil, undergo a degree of "phosphate fixation" with other soil elements or compounds. The degree of fixation depends upon the chemical nature of the soil. The following factors cause phosphate fixation and reduce its availability to the plant.

1. Low soil pH. A low pH increases iron and aluminum either in solution or as precipitates within the soil. In these forms iron and

aluminum can become chemically combined or electrically adsorbed by the ortho-phosphate anion ($PO_4\equiv$). These iron and aluminum phosphates are quite stable and tie up phosphate in forms that make it unavailable to the plant.

2. Addition of too much limestone can result in excesses of calcium, magnesium and carbonates which may combine with and reduce the availability of phosphorous. The ortho-phosphate anion reacts with calcium forming mono, di or tri-calcium phosphate compounds. Phosphorous availability is in direct relationship to the amount of calcium in the molecule. Mono-calcium phosphorous is water-soluble, while di- and tri-calcium phosphorous are not water-soluble. The ortho-phosphate anion can also become unavailable by becoming electrically attached to the carbonates (CO_3) in limestone. It becomes obvious that too little, as well as too much limestone, can be hazardous to the availability of phosphorous.

3. High sodium levels also reduce phosphorous availability. A good soil analysis will reveal potential chemical phosphorous availability problems.

4. In addition to the soil chemical factors listed the clay content of soils can also influence phosphorous availability. As the clay content of soil increases it requires more broadcast phosphorous to maintain a given phosphorous concentration in the soil solution and for crop response. Clay particles are formed of sheets of silica and alumina. It is believed that the ortho-phosphate anion ($PO_4\equiv$) may become electrically attached to the alumina sheets and cause a reduction of water-soluble phosphorous in the soil solution of clay soils.

Occasionally we will not want to build the P2 level of phosphorous. In a soil condition that severely locks it up we would instead place the phosphate in concentrated bands while trying to correct the conditions that tie it up. If the soil test indicates that limestone is needed phosphorous should be banded instead of broadcast until the limestone problem has been corrected. Banding minimizes soil exposure in comparison to broadcasting.

How much phosphorous must be applied to build the soil P2 levels? It requires 4.6 pounds of P_2O_5 to increase the P2 reading by 1 PPM. If a P2 reading was 40 PPM and we wished to raise it to 50 PPM it would have to be increased by 10 PPM. This would require forty-six pounds of P_2O_5 or one-hundred pounds of an 0-46-0 fertilizer. However, applying 4.6 pounds of P_2O_5 would not raise the P1 by 1 PPM because the P1 is never one-hundred percent of the P2. Soil phosphorous levels can be built or maintained either by broadcasting or by banding, depending upon soil conditions.

We will now study the major forms of phosphate fertilizers, including their manufacturing processes, chemical makeup and soil reactions. This information will be used in the material selection and determination of application methods that best fit each particular crop and individual soil conditions.

There are major differences in soil reactions and crop response between the various phosphorous fertilizers. Our study will include the major liquid and dry phosphate fertilizers, starting with rock phosphate, the original source of all of the phosphorous carriers.

ROCK PHOSPHATE

Questions are often raised about the use of rock phosphate to build soil phosphorous levels. At one time this was a very common practice in American agriculture. The decline in the use of rock phosphate does not reflect upon its value as a method of building soil phosphorous levels. Rock phosphate is the source of all of the present-day highly-soluble phosphate fertilizers. Purification and concentration has become the standard of the industry. The first attempt in concentration resulted in normal super-phosphate (0-20-0) which contained about fifty percent calcium sulphate. This in turn was further concentrated into phosphoric acid and triple-super phosphate (0-46-0).

This process does not necessarily improve the product. In many instances the 0-20-0, with its fifty percent load of calcium sulphate, is agronomically superior to the more concentrated forms. Higher concentrations decrease transportation and handling costs, but are not necessarily better products for the soil.

Calcium sulphate is the byproduct caused by converting rock phosphate into phosphoric acid or triple-super. Calcium sulphate is considered to be a neutral liming material and under certain soil conditions can be an excellent soil amendment. This material will be discussed in detail in the **Fertility Section** of the book. The 0-33-0, soft-pebble, Florida Rock Phosphate has been the standard for the industry. There are harder forms that are also good, providing they are properly ground.

Rock phosphate is more difficult to handle and to broadcast than the granular concentrated forms. Transportation costs are much higher. But it is also expensive to process rock phosphate into more concentrated forms. The cost of a unit of P_2O_5 has been less for rock phosphate. All things considered, the rock form is still a very good method of building the phosphorous levels of most soils. The acid action of roots and organic matter decay is sufficient to break down rock phosphate even on relatively alkaline soils.

NORMAL SUPER PHOSPHATE

This product is called 0-20-0 and contains twenty percent phosphorous as water-soluble, mono-calcium phosphorous, plus nine to twelve percent sulfur as gypsum. The calcium sulphate provides a plant-available source of both calcium and sulfur. Its manufacturing process is illustrated in Figure 105A.

ROCK PHOSPHATE	+	SULFURIC ACID (H_2SO_4)	→	$Ca(H_2PO_4)_2$ MONO CALCIUM PHOSPHOROUS 0 - 20 - 0	+	$CaSO_4$ (GYPSUM)

Fig. 105A

Normal super phosphate is produced in a dry, granular form that can be broadcast or used in starter bands. The 0-20-0 contains about seventy-five percent water-soluble (mono-calcium) phosphorous plus calcium and sulphate. It makes an ideal material to apply through grain drills or to use as a starter band for row crops on certain soils that are extremely high in magnesium and low in calcium and sulfur. These soils are found extensively in several of the North Central States.

TRIPLE SUPER PHOSPHATE

This product, called 0-46-0, is made by treating rock phosphate with ortho-phosphoric acid as shown in Figure 105B.

ROCK PHOSPHATE	+	ORTHO PHOSPHORIC ACID (H_3PO_4)	→	$Ca(H_2PO_4)_2$ MONO CALCIUM PHOSPHOROUS 0 - 46 - 0

Fig. 105B

The 0-46-0 contains about eighty percent water soluble mono-calcium phosphorous. It is higher in phosphorous than the 0-20-0 because phosphoric acid is used in processing the rock phosphate instead of sulfuric acid. The sulfuric acid used in the manufacture of 0-20-0 contains no phosphorous so its phosphorous comes entirely from the the rock phosphate source. The 0-46-0 contains only traces of sulfur because there is no sulfur in the phosphoric acid, whereas the 0-20-0 picked up considerable sulfur from the sulfuric acid.

Agronomically, the phosphorous in 0-46-0 is the same as that in 0-20-0. The 0-46-0 is simply more concentrated and without the calcium sulphate byproduct. Both contain water-soluble monocalcium phosphorous (Ca $(H_2PO_4)2$).

AMMONIUM PHOSPHATES

Ammoniated phosphates are formed by combining ammonium and orthophosphoric acid as shown in Figure 105C.

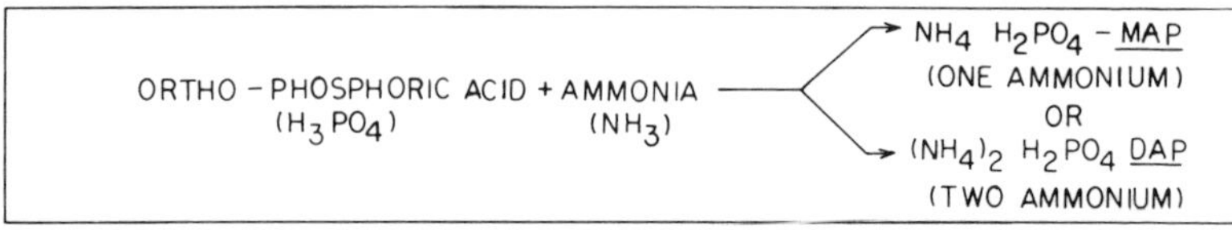

Fig. 105C

The end products are mono- (1) or di- (2) ammonium phosphates, commonly called MAP and DAP. These products contain both nitrogen (NH_4) and phosphorous (H_2PO_4) as shown in Figure 105D.

MAP = 11 - 48 - 0	$NH_4H_2PO_4$	(ONE AMMONIUM)
DAP = 18 - 46 - 0	$(NH_4)_2$ H_2PO_4	(TWO AMMONIUM)

Fig. 105D

These two forms of ammonium phosphates are very similar with the exception being that MAP is acid and DAP is alkaline. The concentrated solution at the granule site in the soil may have a 4.0 pH for MAP and up to a 9.0 pH for the DAP.

DAP would tie up faster as di-calcium or tri-calcium on an alkaline soil (high in calcium) than MAP. MAP should be selected for broadcast on high pH or alkaline soils and DAP used for soils that are on the acid side. Everything possible must be done to minimize soil "lock up" of phosphorous.

Placement can also be a determining factor in the selection of either MAP or DAP. High concentrations of ammonium can be damaging to germinating seeds and roots. For banding applications close to the row we prefer to use MAP instead of DAP. Under certain conditions the high concentrations of ammonium in DAP can cause roots to burn instead of proliferating.

If the choice is available we will always recommend MAP instead of DAP for the purpose of banding near the row. We want good phosphorous uptake and rapid root growth early in the life of the plant, so why take a chance with DAP which could damage roots? Banding of MAP is therefore preferred to DAP even on acid soils.

DAP may be used in banding applications if it can be placed deeper and farther from the row than normal. By placing DAP three inches or more to the side and possibly deeper than the usual one and one-half inches below the seed there is more time for the ammonium to dissipate before the roots reach the band. Dilution is the key to prevention of root damage from excessive ammonium. DAP can be safely broadcast without fear of root damage because the ammonium is not as concentrated as in a band.

Normal super (0-20-0), triple-super (0-46-0) and ammoniated forms of phosphates (MAP and DAP) are the four major manufactured dry forms of mono-calcium ortho phosphate fertilizers used in agriculture.

ORTHOPHOSPHORIC ACID

Orthophosphoric acid (H_3PO_4) is a liquid and a constituent of most liquid or dry phosphate fertilizers. Both normal and triple super phosphate contain water-soluble mono-calcium orthophosphoric acid represented by the formula $Ca(H_2PO_4)2$. MAP and DAP contain mono- or di-ammoniated orthophosphoric acid. The orthophosphoric acid molecule (H_3PO_4) contains only one phosphorous atom and represents the form most readily available to the growing plant.

Two processes are used in the manufacture of orthophosphoric acid:

Electric Furnace Process. This method produces a clear, water-like, "furnace acid" product that is quite free of impurities and is the source of most clear liquid phosphorous fertilizers. The general manufacturing process is shown in Figure 105E.

ROCK PHOSPHATE + HEAT ⟶ P (PHOSPHOROUS) + O_2 ⟶ P_2O_5 + H_2O ⟶ H_3PO_4 (ORTHO PHOSPHORIC ACID)

Fig. 105E

In this heat process, finely ground rock phosphate is heated to drive off elemental phosphorous. The elemental phosphorous combines with oxygen to form P_2O_5. The P_2O_5 is then combined with water to produce the furnace grade, clear liquid, orthophosphoric acid.

Wet Process. Instead of heating the rock phosphate, it is treated with liquid sulfuric acid or phosphoric acid as shown in Figure 105F. The final liquid must be filtered to remove the many impurities, such as calcium sulphate, from the rock phosphate source.

Both of these processes produce liquid orthophosphoric acid containing fifty-two to fifty-four percent phosphorous.

ROCK PHOSPHATE + SULFURIC ACID OR PHOSPHORIC ACID → H_3PO_4 (ORTHO PHOSPHORIC ACID) + IMPURITIES

Fig. 105F

SUPER PHOSPHORIC ACID

As we learned earlier, orthophosphoric acid is the form of phosphorous used in most phosphate fertilizers and is the form most readily available for plant uptake.

The ortho form is most reactive to other elements or compounds in the soil and can quickly become "tied up." Orthophosphate availability can be greatly reduced within hours of application. It can react with most of the basic elements such as calcium, magnesium, sodium, potassium iron and aluminum and become non-water soluble. As a result of these reactions its availability may be reduced from slight to nearly one hundred percent. Maintenance of availability is the goal in phosphorous management.

Superphosphoric acid is most commonly call **polyphosphate.** The polyphosphate, or super-phosphoric acid, is produced by condensing or dehydrating the ortho-phosphoric acid. Depending upon the amount of water removed, super-phosphoric acid will contain forty to seventy-five percent polyphosphate and twenty-five to sixty percent ortho-phosphate. The differences between the ortho- and polyphosphate chemistry are quite important as they relate to soil reactions affecting phosphorous availability.

As outlined earlier, orthophosphate has only one phosphorous molecule in its formula, is extremely reactive with other soil elements and may become quite unavailable within hours after being applied to the soil. Polyphosphate, on the other hand, is a more stable compound, normally containing two to ten phosphorous atoms chemically bonded together.

Figures 106, 107 and 108 are molecular drawings of the orthophosphate, orthophosphoric acid and polyphosphoric acid molecules. Notice that the orthophosphate (Figure 106) contains only one phosphorous atom connected to four oxygen atoms (PO_4). Ortho-, as such, is not an acid until it hooks up with three hydrogen atoms (Figure 107), forming orthophosphoric acid (H_3PO_4). Orthophosphoric acid still has only one phosphorous atom. The polyphosphoric acid molecule (Figure 108) contains more than one phosphorous atom, hence the name **poly.** It is still a phosphoric acid because it contains hydrogen, but it is now a poly-(more than one phosphorous)

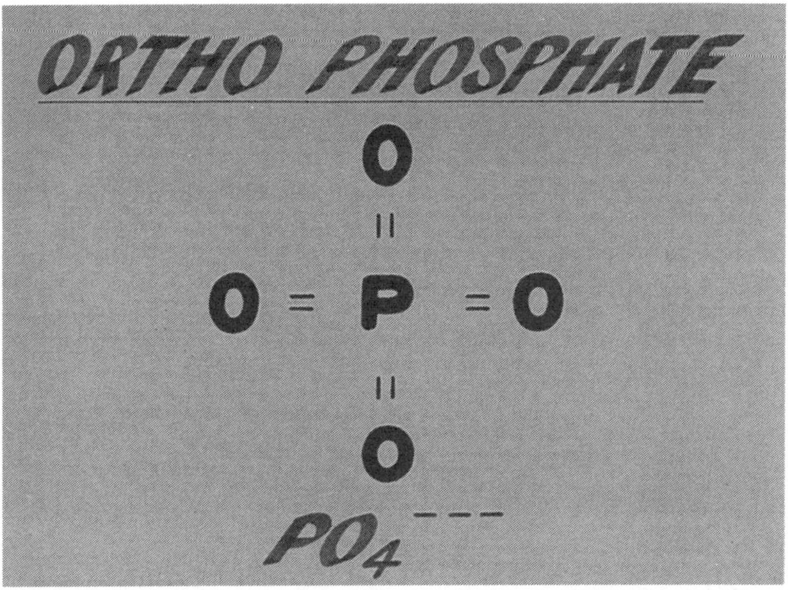

Fig. 106

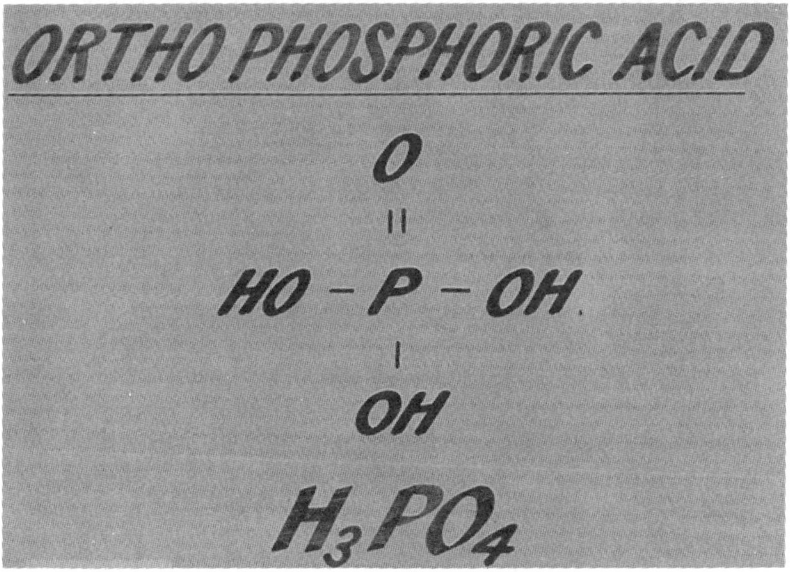

Fig. 107

phosphoric acid. The poly form (Figure 108) is quite stable unlike the ortho- form and will not tie up with other soil chemistry.

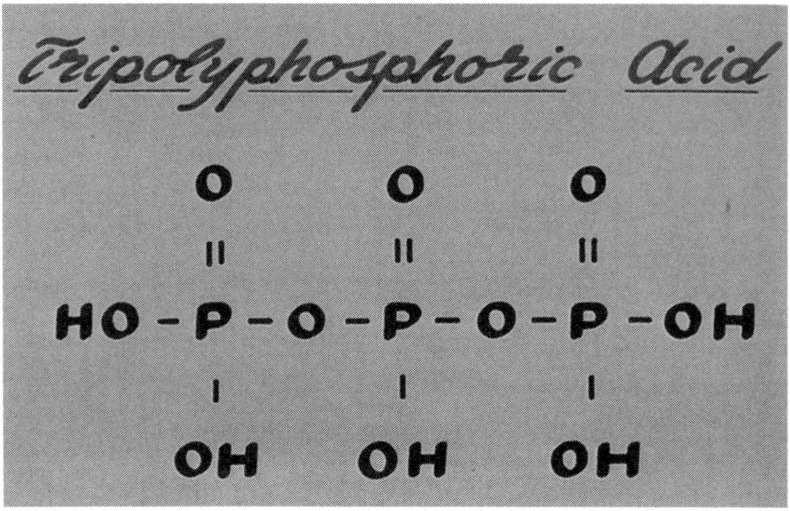

Fig. 108

When applied to the soil the poly form will change back to the orthophosphate form within a period of a few days or weeks depending upon soil conditions. The soil factors favoring soil microbiological activity also speed the conversion of poly back into the ortho form.

Although it is generally concluded that plants can utilize some polyphosphate, the hydrolysis of polyphosphate back into the ortho form is necessary for the most efficient phosphorous recovery.

Polyphosphates, being less reactive with other soil minerals than the ortho forms, maintain phosphorous in a protected position slowly releasing it to the growing plant.

The period of time required for poly to convert back to the more available ortho form is sufficient to permit adequate root development to begin the important process of phosphorous recovery. The early protection from phosphorous fixation and its timely release are important traits which make polyphosphate a very desirable source of phosphorous.

The superphosphoric acid (polyphosphate) always contains some orthophosphate. The combination is excellent.

Notice in Figure 109 how the orthophosphate form expands outward from the fertilizer band leaving the polyphosphate remaining within the band. The polyphosphate remains mostly in place until it changes to the ortho form. The advancing wall of orthophosphate, expanding out from the fertilizer band, combines with calcium, magnesium iron, aluminum, etc., thus neutralizing the soil's phosphorous fixation properties within the expansion zone.

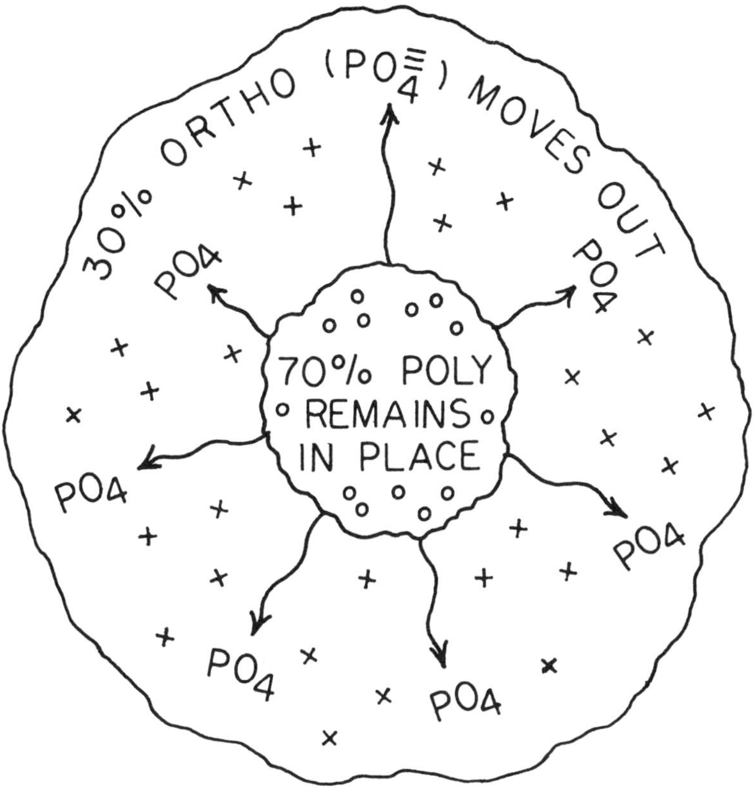

Fig. 109

As the polyphosphate converts back into orthophosphate, it will also move outward, traveling through the now relatively phosphate fixation-free zone before it encounters lock-up conditions. The neutralizing effect of the orthophosphate in the expansion band, combined with the slow or delayed release of the poly form, combine to assure phosphorous availability by the time the roots reach the fertilizer band.

Maintaining phosphorous availability is an important consideration in overall fertility planning. The polyphosphate which contains some orthophosphate provides a combination which insures long-term availability. Polyphosphate is an ideal form of phosphorous for use as "pop up" or in starter bands. Its use as such will be thoroughly covered in the **Fertility Systems** section of this book.

AMMONIUM POLYPHOSPHATE

Super phosphoric acid containing at least fifty percent polyphosphate can be combined with ammonia to form the ammoniated polyphosphate liquids 10-34-0 or 11-37-0. The 10-34-0 is usually produced from ammoniating super **green** acid while the 11-37-0 is made from the super **white** acid. The manufacturing of the ammoniated polyphosphate solutions 10-34-0 and 11-37-0 is called a "hot mix process" because heat is produced in the mixing of ammonia and superphosphoric acid.

However, it is important to understand that a 10-34-0 blend can also be made from non-poly sources. A given analysis can be made from almost any combination of products and the analysis has little bearing on product quality.

As with super phosphoric acid, the ammoniated polyphosphate (10-34-0 or 11-37-0) contains both ortho and poly phosphates. The soil reaction is identical. The ortho fraction will expand and serve as a neutralizing front, as shown in Figure 109.

With a basic understanding of phosphorous chemistry it will be easier to choose materials to best fit your own soil conditions. You must specify to your fertilizer dealer the exact form you want, such as poly, MAP, DAP, etc. The formula, as such, is not always a guarantee of form or quality of material.

POTASSIUM

Potassium (K) figures are given in PPM per acre. K is the symbol for potassium and will be used in the subsequent discussion. Potash refers to potassium fertilizer materials.

There are three major potassium fertilizers used in the United States.

1. **Potassium chloride** (KCl^-), commonly called muriate of potash (0-0-60). Muriate contains sixty percent K_2O or fifty percent elemental potassium. To convert K_2O to elemental potassium it must be divided by 1.2. Crop needs, however, are usually given as K_2O.

Potassium chloride is a salt because it contains chloride just as sodium chloride (table salt) does. It can be either pink, red, or white in color depending on the mining or recovery process. There is no measurable difference in their agronomic values.

Keep in mind that 0-0-60 is a salt, and as such it can be harmful to roots and even to soil life. This salt factor must be taken into consideration if it is to be used in starter fertilizer bands.

When applied to the soil 0-0-60 dissolves in soil moisture and breaks apart as K+ and Cl^-. The K+ can adhere to the negative soil particle and the chlorine (Cl^-) will leach out or combine with calcium and form calcium chloride or road salt.

Muriate of potash (0-0-60) is the most widely used source of potassium wth the exception of tobacco or potatoes. These two crops, expecially tobacco, are very sensitive to the chloride. Potato growers use 0-0-60, but should not use it in excessive amounts in starter bands or pre-plant applications. Fall applications permit the chloride to dilute or rinse out of the soil's root zone.

2. **Potassium sulphate,** or sulphate of potash (K_2SO_4). This product contains fifty percent K_2O (0-0-50) plus seventeen percent sulfur in the ready-to-use sulphate ($SO_4=$) form. This is an excellent product and is in heavy demand by the tobacco growing industry. Potato growers would do well to use this product instead of 0-0-60. Potassium sulphate is quite a bit more expensive than 0-0-60 until one figures in the value of the sulfur. One hundred pounds of 0-0-50 per acre annually would maintain the soil sulfur levels. If sulfur is needed its value must be considered in the product cost. After subtracting the value of the sulfur the unit cost of the K_2O can then be compared.

3. **Potassium magnesium sulphate,** or sul-po-mag. Sul-po-mag contains twenty-one percent K_2O, eleven percent magnesium and twenty-two percent sulfur. The potassium, magnesium and sulfur are all in the sulphate form.

Sul-po-mag is normally used on soils that are deficient in magnesium and cannot be corrected with limestone because of excessive calcium levels. Sul-po-mag is an excellent source of sulfur. It is much lower in potassium than the other two sources. If one is opposed to the use of chlorides in the soil the 0-0-50 would be a more economical source of potassium unless magnesium is needed. Sul-po-mag can be used when magnesium is needed. The sulfur and potassium can then be considered a bonus. It is not economical to use sul-po-mag as a source of potassium and sulfur when the soil magnesium levels are high.

Potassium is a very important plant nutrient. Some of its functions are:

a. Producing or regulating the activities of forty or more enzymes;
b. Being responsible for cellulose production and cell wall strength which results in resistance to disease and lodging;
c. Increasing oxygen intake through stomata openings, thus increasing photosynthesis and production of sugars;
d. Facilitating the movement of sugars and starches to sites of need. Sugar movement is important to fill grain and to prevent backlogging in chloroplasts which causes a reduction of photosynthesis;
e. Playing an important role in the conversion of nitrogen into protein and thereby reducing non-protein nitrogen;
f. Reducing respiration, thereby leaving more energy for production; and
g. Improving drought resistance.

It can readily be seen that potassium has an involvement with nearly every aspect of plant growth. Potassium "makes things happen" within the cells without actually becoming an integral part of the protoplasm or cell structures. It serves mainly as a catalyst. Because of this loose existence potassium is easily leached from dead plant tissues. Cornstalks and small grain straw will leach a considerable amount of their potassium between harvest and spring. Potassium is also easily leached or washed out of manure piles exposed to the weather. Rain can leach potassium from freshly mown hay.

Potassium is a major element. The amount taken in by plants is exceeded only by nitrogen. The uptake of potassium is critical during early stages of growth.

Tests done in Ohio, for example, showed that corn took up about seventy percent of its total potassium needs within the first sixty days of growth. This is a much higher percentage of potassium than of either phosphorous or nitrogen. This is why it is so important to assure high potassium availability, especially during the periods of greatest intake.

Plants use potassium as elemental $K+$. Its availability to the plant depends upon its position within the soil in relationship to clay, humus and soil water.

Potassium can be found in the following positions within the soil structure:

1. **Unavailable potassium.** Most potassium has its origin from the potassium-bearing minerals feldspar, muscovite and biotite. All three minerals contain potassium (K) within their chemical structures, but in all three it is locked within the chemical structure and remains unavailable to the plant. Once these minerals are physically broken down into fine powder soil acids can dissolve out the individual elements in a process called "chemical weathering." This is an extremely slow process and under most conditions it will not release enough potassium to maintain soil levels. The amount of these minerals forming a soil's parent material determines a soil's reservoir, or "native potassium supply."

Certain soils, such as those in the Platte River Valley, are extremely high in native potash because of the windblown, potassium-bearing, granite dust that was blown in from Canada during the last glacial period.

2. **Slowly available potassium.** Clay particles are unique because they are extremely small and each tiny particle contains two or three distinct layers. Clay particles are similar to tiny accordians because they can expand and contract as they take on or lose water. As water molecules move between the clay layers they expand, permitting potassium (K) to move in between the layers. This is called "store-

house" potassium. The potassium positioned between these layers will not be available to the roots until it works its way back to the exterior surface of the clay layers. When the clay particles lose water the layers contract, trapping the potassium until water again expands the layer.

3. **Readily available potassium.** This source of potassium is held within the soil in two ways. First, it is held on the negative cation exchange sites. Since potassium has a positive charge it is attracted to a negative site on the soil colloid and is held like a nail to a magnet. Roots can remove this potassium directly from the surface of soil colloids. The second way in which readily available potassium is held is in the water film around the soil particles. This is the most readily available form and these potassium rich moisture films are the primary source of potassium for the plants.

The exchangeable potassium held on the soil particles can diffuse from the colloidal surface into the surrounding moisture film. This is a diffusion or equalization process whereby the exchangeable potassium tends to remain in equilibrium with the potassium in the surrounding moisture film. This does not mean in exactly equal amounts however. To be in equilibrium it is estimated that less than ten percent of the exchangeable potassium will be in solution.

As the exchangeable surface potassium is depleted the "storehouse" K will move very slowly out from between the clay layers back to the surface exchange sites in an attempt to again reach and maintain an equilibrium.

We can readily see how dry weather affects potassium availability. Drought can induce a potassium deficiency by reducing the K in solution and trapping it between the clay layers as they contract from loss of water.

Figure 110 illustrates the entire process. The exchangeable and "storehouse" potassium can exchange positions as shown in the "B" exchange. Exchange "A" shows that the potassium in solution and on the exchange sites can move in either direction to maintain equilibrium. **When water-soluble potash fertilizer is applied the soil solution becomes highly concentrated and the movement of potassium is from the solution to the surface exchange sites.**

A clay particle is a very strong magnet in comparison to sand, silt and humus. Clay soils hold potassium very tightly and resist leaching. This trait also makes it more difficult to recover potassium from clay soils.

Sand, silt and organic soils will lose more potassium through leaching while soils high in clay and with a normal soil nutrient balance will leach very little potassium through tile water. **A soil's**

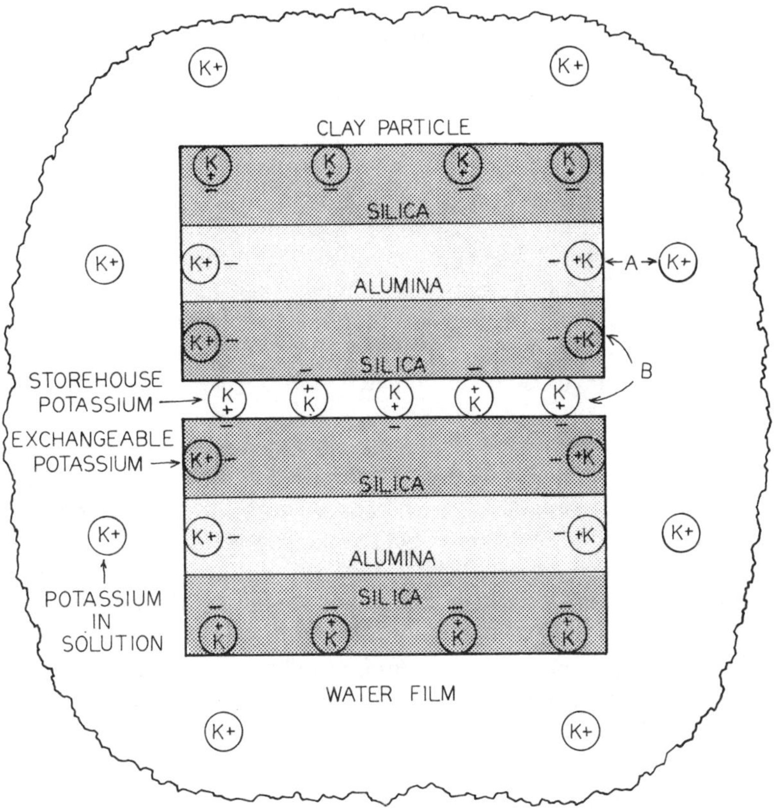

Fig. 110

nutrient-holding capacity and leaching characteristics will certainly influence both the rates and timing of fertility applications.

We now address the problem of making potash recommendations to meet the demands of each particular crop. Several factors must be considered including the type of crop being grown and maintenance of potassium levels for adequate production in relation to each soil's cation exchange capacity.

Figure 111 offers a suggested guideline for potassium levels in PPM and based on the soil's CEC. These levels would provide a high to very high soil test reading for potassium.

These are the soil levels we should work toward. As a guideline, it will require four pounds of 0-0-60 (2.4 pounds K_2O) to raise soil potassium levels by 1 PPM. But, this will not hold true for all soils and certain clay soils may require eight to twelve pounds of potash to increase the soil potassium level by 1 PPM because of K being

CEC NUMBER	SOIL TEST K LEVELS
5 or Less	125 PPM
6 to 10	150 PPM
FOR EACH CEC OVER 10	ADD 10 PPM

Fig. 111

absorbed between the clay layers. A clay soil that has become deficient in potassium will absorb more potassium between the layers than one that has been well fertilized. As clay soils become deficient in potassium the storehouse K also becomes deficient as it works its way out to maintain an equilibrium with the surface or exchangeable potassium.

Each soil's cation exchange capacity is an important consideration because low CEC soils hold fewer cations, but release them, including potassium, easier than high CEC soils. Low CEC soils are also more subject to leaching of cations.

A 3 CEC soil at a five percent K base saturation would contain 117 pounds K per acre, or the equivalent of 141 pounds of K_2O. The 125 PPM, as recommended in Figure 111 for a CEC under 5 would equate to a 10% potassium (K) base saturation for a 3 CEC soil.

It requires about 190 pounds of K_2O to grow 150 bushels of corn. A 3 CEC soil, at 5% base saturation and containing 140 pounds of K_2O, would not have enough potassium to grow 150-bushel corn even if one hundred percent of the soil potassium was recovered. The PPM recommendation is a better guide than the percent base saturation figures, particularly on low CEC soils. A degree of potassium "flooding" of the soil is necessary to sustain high yields on a low CEC soil.

If a soil is oversaturated with a cation, such as potassium, leaching losses are also increased. If a cation cannot attach to a negative site it will leach downward with the soil water.

Timing of the potash application is important on low CEC soils. If, in order to produce high yields, we must apply more potassium than the soils can hold, it should be done in the spring rather than the fall to shorten leaching time.

If it is feasible, banding would be superior to broadcasting on low CEC soils. Banding would provide zones of high potassium concentration and minimize competition for exchange sites from other cations such as calcium and magnesium. Potash bands, spaced eighteen to twenty-four inches apart, could be applied on the surface or ahead of seedbed tillage on sandy soils. The potassium would penetrate adequately because of the low holding capacity of sands. Surface-banding of a cation such as potassium would not be done on a clay soil because it would not penetrate sufficiently into the root zone.

The type of crop to be grown will also influence potash applications. Grasses, such as corn, can remove potassium much more efficiently than legumes. Topdressing legumes is a better method of providing additional concentrations of potassium instead of building excessive soil levels. Topdressing is a method of "spoon feeding" a crop without causing a cation imbalance.

We must strive to maintain a cation balance within the soil. There are a given number of negative sites available to hold cations and, if we oversupply one cation, it is at the expense of another. Balancing potassium as outlined in Figure 111 will be adequate for most crops.

Soils that test low in potassium can be "built up" over a period of years without sacrificing yields. Spring broadcasting is normally more efficient than a fall application because of the reduction in leaching time on low exchange soils and reducing "fixation time" on clay soils. This is an important consideration during the buildup phase or when resources are limited.

During "buildup," banding fifteen to thirty pounds of K_2O per acre in a normally placed starter fertilizer band can provide good insurance without fear of excessive salts near the row. Even higher rates can be used if the starter band is placed farther from the row, as outlined in the unit on **Developing a Fertility Program.**

Soil magnesium levels must be checked in making potassium recommendations. High potash applications on soils that are marginal in magnesium can cause a magnesium deficiency in the crop.

The postassium to magnesium ratio appears to be quite important. For a **normal** magnesium range of twelve to fifteen percent base saturation the postassium saturation should be about **one-third** of the magnesium saturation. This would not be practical for soils with twenty to thirty percent magnesium saturation. With such extreme excesses it might be beneficial to band additional potassium to assure a balanced uptake between the two cations. The importance of the potassium to magnesium ration emphasizes the importance of nutrient balance.

High sodium soils will affect potassium application rates. The sodium and potassium elements are so similar that the plant will actually take up sodium as if it were potassium. Since sodium does not perform all of the functions of potassium its uptake in the place of potassium can greatly reduce plant performance.

We believe that whenever the base saturation of sodium reaches **one-third** or more of the base saturation of potassium a deficiency of potassium must be assumed. There can be sufficient substitution of sodium for potassium to cause a degree of potassium deficiency.

Under these conditions the banding of potash is a very profitable practice. The high concentration of potassium within the band increases potassium availability and uptake, thus reducing the substitution of sodium.

In summary, we conclude that recovery of potassium as with any other cation is a matter of concentration of the available nutrient in relationship to the root mass. Soil aeration is essential to the nutrient recovery process.

MAGNESIUM, CALCIUM & SOIL pH

Magnesium, calcium and soil pH will be discussed singularly and also as a unit because all three relate to the selection and application of liming materials. Liming is an important consideration in a total soil fertility program, but it is often abused simply because the reasons for applications of limestone are not fully understood.

In this unit of study we will show how limestone adjusts the soil pH and how to select and apply a liming material based on the balancing of the cations, calcium and magnesium.

The soil test gives a calcium and magnesium readout in PPM as well as in their percent base saturation (pantry fill). These figures and those of the soil pH and CEC will be used in determining the type of liming material to be used as well as its rate of application.

SOIL pH

Our study of soil pH begins with the meaning and significance of soil pH. Figure 112 illustrates a pH scale as it relates to the soil. The

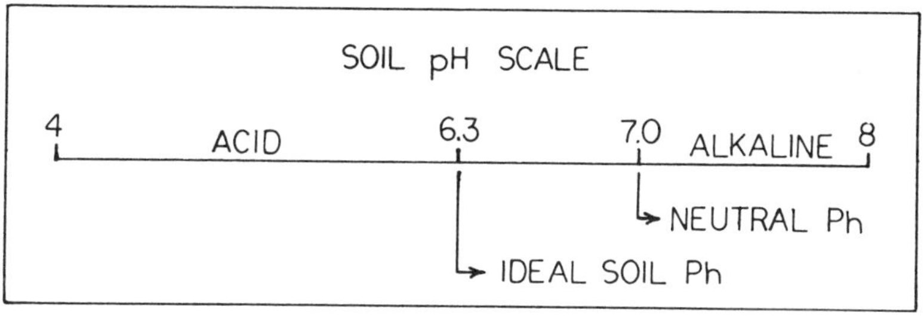

Fig. 112

scale begins with a low pH of 4.0 and goes up to a pH of 8.0. Actually, a soil may have a pH either below or above these figures.

A pH of 7.0 is neutral. As the soil pH goes below 7.0 it becomes increasingly acid and it becomes more alkaline as it goes above the neutral 7.0. Soil pH up to a neutral 7.0 is a measurement of hydrogen. The amount of hydrogen in the soil determines the acidity of the soil.

Hydrogen is the acid element. The scale shows a 6.3 soil pH which is considered to be near ideal for most crops. At a 6.3 soil pH the soil test will usually show a base saturation, or pantry fill, of ten to eleven percent hydrogen.

At the neutral pH of 7.0 there is no measurable hydrogen on the colloid. As the soil pH goes above the neutral 7.0 hydrogen remains at zero, but other alkaline elements such as sodium, calcium or magnesium could increase. Below the neutral 7.0 soil pH measures only hydrogen which is not even a plant nutrient. By studying your soil test results you will observe how the soil pH goes **down** as hydrogen percent base saturation **increases.**

Figures 113 and 113A will give a good mental picture of soil pH. Figure 113 shows an acid soil particle containing hydrogen with a pH below the neutral 7.0. Figure 113A shows a soil with a neutral 7.0 pH. At a pH of 7.0 there is no hydrogen on the soil colloid.

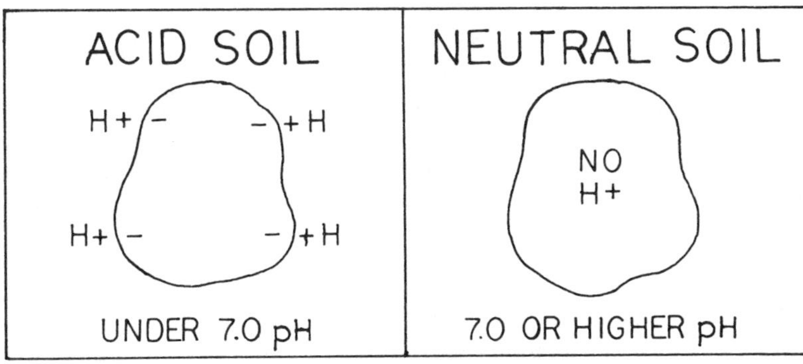

Fig. 113 **Fig. 113A**

The soil colloid will **always** be completely filled with cations. **When cations are removed and not replaced hydrogen moves in to fill the negative sites once occupied by the other cations as calcium, magnesium and potassium.** As calcium, magnesium and potassium leave the soil colloid hydrogen fills the void and the soil becomes increasingly acid and less productive. We believe crop production failure is caused more because of a lack of nutrients rather than because the soil pH becomes lower.

As plant nutrients are removed from the soil colloid hydrogen continues to fill the void until the cupboard is bare of plant nutrients. Try to visualize a soil pantry nearly filled with the non-plant nutrient hydrogen and practically empty of the essential cationic plant nutrients, calcium, magnesium, potassium, etc.

This whole relationship between soil pH, hydrogen, calcium and magnesium is important because it relates directly to the correct selection of limestone, its rate of applcation and the frequency of application. Instead of applying limestone only to correct the soil's pH its application must provide the essential plant nutrients calcium and magnesium.

LIMESTONE CHEMISTRY AND SOIL REACTIONS

Applications of limestone demand a knowledge of the different types and how each reacts chemically within the soil to alter the soil pH, calcium and magnesium.

Basically, limestone contains calcium or calcium plus magnesium. Rates of application and selection of limestone type must be based on soil test data that indicates a need to adjust calcium or both calcium and magnesium. Limestone should never be applied unless a soil analysis indicates a specific need. Limestone must not be applied indiscriminately, and above all, those who preach that you cannot apply too much limestone must not be believed. Recommendations as high as ten to fifteen tons per acre of a specific limestone per acre have even been made, often without the benefit of a soil analysis for a guide.

Three basic types of liming materials must be examined because, depending upon your soil analysis, a selection must be made between them whenever you apply limestone.

1. **Calcium limestone.** This liming material may also be called "calcite," "calcareous," "high-calcium lime," or "calcium carbonate lime." It contains mostly calcium carbonate ($CaCO_3$) and has little, if any, magnesium. Calcium is the only cation it contains and is the reason it is often referred to as "high-calcium limestone." Pure calcium carbonate contains forty percent of calcium. It would be rare to find pure calcium carbonate, but for practical purposes the mined calcium carbonate can be figured at forty percent calcium.

There are several sources of calcium carbonate limestone. It is available as regular agricultural limestone in several parts of the country. Other sources include sugar beet lime, paper mill sludge and the byproducts from city water treatment systems. Sugar beet lime results from using powdered calcium carbonate to purify the sugar solutions during sugar beet processing. The same lime is used in processing wood pulp into paper. Water treatment plants use this lime in their purification systems. These byproducts carry impurities usually of an organic nature, that are harmless to soils and plants. As a result of their impurities their calcium levels are lower than regular limestone. They are in a powdered form and are excellent sources of calcium. Most byproduct liming materials are calcium carbonate whether from the beet, paper mill or water processing industries and are good sources of high calcium limestone.

Another source of calcium is marl. Marl deposits are usually found in association with swamps or bogs and can be an excellent source of calcium cabonate. Marl must be mined and allowed to dry before it can be conditioned for spreading.

2. **Dolomite limestone.** Dolomite limestone contains both calcium carbonate ($CaCo_3$) and magnesium carbonate ($MgCO_3$). It always contains more calcium than magnesium. While all limestones contain calcium they do not all contain magnesium. Any limestone which contains magnesium is referred to as "dolomitic limestone."

Regular mined limestone always contains its calcium and magnesium in the carbonate forms. This will be important to remember when studying limestone reactions when applied to the soil. Pure calcium carbonate contains **forty** percent calcium and pure magnesium carbonate contains **twenty-eight** percent magnesium.

Limestones are often rated according to their neutralizing powers. Magnesium carbonate will neutralize more soil acid than an equal amount of calcium carbonate, so dolomitic limestone will always have a higher neutralizing rating than calcium carbonate limestone. Calcium carbonate limestones are assigned a neutralizing rating of one hundred. A dolomitic limestone will be rated over one hundred, depending on how much magnesium cabonate it contains. Magnesium carbonate will neutralize 1.19 times the soil acidity of an equal weight of calcium carbonate.

The term, dolomite, does not infer a set level of magnesium carbonate. Dolomite limestones can vary from less than five percent to more than thirty percent magnesium carbonate. Regardless of the level it is all called dolomitic limestone.

While you should be aware of how the neutralizing values of limestone are derived, if applying limestone for either calcium alone or calcium and magnesium, you will seldom be concerned with buying limestone because of its neutralizing powers alone. It would be a serious mistake to purchase dolomitic limestone just becasue it had a lower cost to apply to soils already too high in magnesium.

3. **Calcium sulphate.** Calcium sulphate ($CaSO_4$) is often called "gypsum." In some areas of the country it is also called "land plaster." Calcium sulphate contains both calcium and sulfur. The sulfur is in the sulphate form. You will note that **calcium sulphate does not contain any carbonates** such as the other two limestone forms. This lack of carbonates is important when we refer to gypsum as a neutral liming material. Gypsum is a specialty product used in solving a variety of soil problems. It normally contains **twenty-two** percent calcium and **eighteen** percent sulfur.

We will limit this study to the above three types of liming materials although there are other sources, such as calcium oxide (CaO) and magnesium oxide (MgO), which are formed from heating calcium and magnesium carbonates. Heat simply drives off carbon dioxide leaving calcium oxide, which is very high in calcium and quite caustic. Calcium oxide can burn seedlings so it must be handled accordingly.

Limestone influences soil chemistry by changing the soil pH and either calcium by itself or both calcium and magnesium levels. The impact of these three basic changes will influence many other soil chemical and biological functions. Our concern is simply to set the stage for good overall soil fertility and plant nutrition by intelligently managing the liming phase. This demands a basic understanding of how limestone reacts to bring about desirable changes in soil chemistry.

To keep the matter simple, let us review a soil particle as shown in Figure 115. You will recognize the soil particle as having a number of negative sites that attract positive-charged cations. In the example we will limit the cations to calcium (Ca), magnesium (Mg), potassium (K) and hydrogen (H). Remember, every negative site will be satisfied with a positive-charged cation. And, as calcium, magnesium and potassium are depleted from the soil particle (pantry), hydrogen (H+) will move in to occupy the negative sites as shown in Figures 116 and 117. You will recall that hydrogen is not a plant nutrient, but an acid element.

Fig. 115 **Fig. 116** **Fig. 117**

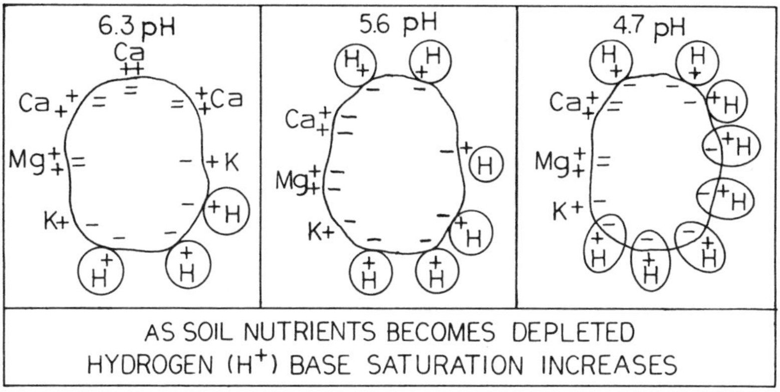

AS SOIL NUTRIENTS BECOMES DEPLETED
HYDROGEN (H^+) BASE SATURATION INCREASES

Soil acidity increases as the hydrogen increases on the soil particles. Notice the progressive lowering of the pH from 6.3 to 5.6 and on to 4.7 as hydrogen replaces calcium, magnesium and potassium as shown in figures 115, 116 and 117. This effectively demonstrates why crops fail because of the lowering of soil nutrients rather than the lowering of the soil pH alone.

When limestone is added this trend is reversed by increasing the calcium, magnesium and soil pH. Obviously, if limestone raised the soil pH, it must also lower and remove the acid element, hydrogen. If hydrogen is removed the negative soil sites are exposed making room for other cations such as calcium and magnesium. This is exactly what liming accomplishes.

Figure 117A shows what happens when limestone is applied to the soil. To simplify our explanation we used calcitic, or calcium carbonate limestone ($CaCO_3$), but the magnesium carbonate ($MgCO_3$) will work in exactly the same way.

$$CaCO_3 \xrightarrow[\text{ACIDS}]{\text{SOIL}} Ca^{++} + CO_3^{=} \text{ (CARBONATE)}$$

Fig. 117A

When calcium carbonate limestone is applied the soil acids break the molecule into its separate components of calcium (Ca++) and carbonate (CO_3=). Notice that calcium carries two positive charges and the carbonate two negative charges.

In this reaction we have one calcium and one carbonate. This can now be viewed as a soil reaction in Figures 118 and 119. Figure 118 shows a soil particle with three hydrogen atoms on it. This soil would

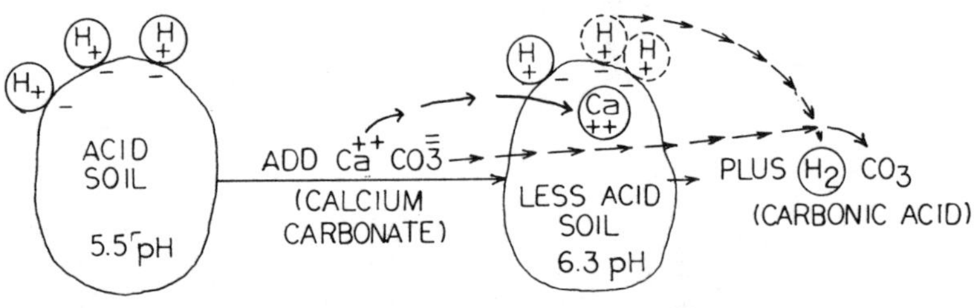

Fig. 118 **Fig. 119**

be acid. Figure 119 shows the same soil with a calcium molecule and one hydrogen. By adding one calcium (Ca++) and one carbonate (CO_3=), we can remove the two hydrogens and replace them with one calcium.

How did this happen? First, we must understand that a carbonate is very reactive and will not stand alone. A carbonate quickly reacts with hydrogen and since carbonate has two negative charges to satisfy it will attract two positive charged hydrogen atoms as shown in Figure 119A.

The carbonate and hydrogen combine to form carbonic acid. The carbonate captured two hydrogen atoms from the soil particle, leaving

$$CO_3^{=} + 2H^{+} \longrightarrow H_2CO_3$$

(CARBONATE) (CARBONIC ACID)

Fig. 119A

two negative sites exposed. A calcium with two positive charges (Ca++) can then occupy these two exposed negative sites (Figure 119).

This reaction raises the soil pH from 5.5 in Figure 118 to 6.3 in Figure 119 because two acid hydrogen elements were removed from the pantry. One calcium moved into the pantry to replace the two hydrogen. In this example the added limestone (calcium carbonate) raised the soil pH by removing two hydrogen, but more importantly, it increased the soil calcium levels. If dolomitic limestone had been applied both calcium and magnesium would have been increased and the soil pH raised as well.

The purpose of this illustration is to show the basic functions of limestone so we can think clearly about applications of limestone in terms of increasing calcium, or calcium plus magnesium, and not simply to adjust the soil pH. Understanding this principle will make you very careful in selecting the correct type of limestone to fit your soil needs which must go beyond the consideration of pH alone.

Knowing how the carbonate limestones work will make it much easier to understand the function of a third liming material, calcium sulphate ($CaSO_4$). When calcium sulphate is applied, as shown in Figure 119B, soil acids break the molecule into its separate component parts of calcium (Ca++) and sulphate (SO_4=). In this reaction

$$Ca\,SO_4 \xrightarrow[\text{ACIDS}]{\text{SOIL}} Ca^{++} + SO_4^{=}$$

(GYPSUM) (SULPHATE)

Fig. 119B

we have one calcium and one sulphate (Figures 120 and 121). Note that calcium sulphate differs from limestone because it contains sulphate instead of carbonates.

Figure 120 shows a soil particle with two hydrogen, two magnesium (Mg++) and two sodium (Na+) atoms.

Figure 121 shows the same soil particle still containing the two hydrogen and two sodium atoms, but only one magnesium. Note that calcium has replaced the magnesium. The sulphate (SO_4=) combined with the magnesium (Mg++), forming an epsom salt ($MgSO_4$)

crystal. Since magnesium was removed from the soil the calcium from the calcium sulphate can move in to replace it.

What happens to the soil pH? The sulphate does not remove hydrogen as the carbonate in limestone does. Since hydrogen remains the same, the soil pH also remains the same. Remember, hydrogen is the acid element.

Sodium (Na) is also shown on the soil particles in Figures 120 and 121. Sulphates can also combine with sodium, removing it from the

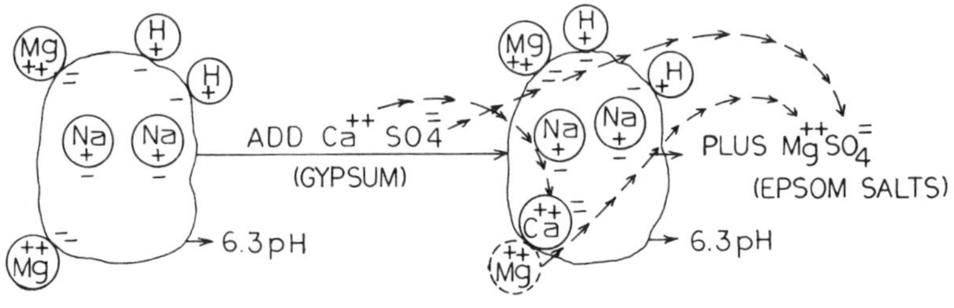

Fig. 120 **Fig. 121**

pantry. Therefore, calcium sulphate can serve several soil functions. It can reduce sodium and magnesium as well as supplying calcium and sulfur. Used in sufficient amounts calcium sulphate can help overcome excessive sodium or magnesium.

Calcium sulphate (gypsum) is often referred to as a "soil conditioner" because it often brings about a mellowing effect on heavier soils. The improvement in soil tilth is probably brought about because of increased biological activity due to the increased calcium and sulfur. The lowering of magnesium accompanied by an increase in calcium can improve the tilth on heavy soils. High sodium causes soils to become very tight or compacted. Calcium has a flocculating or granulating effect on clay colloids. Reducing the sodium will greatly improve the soil structure.

But, farmers are still being sold calcium sulphate as a soil conditioner without the benefit of a soil test. The use of gypsum on a soil that is low in magnesium could be disastrous by lowering it further. Neither would gypsum normally be used strictly for calcium or sulfur because there are much less expensive sources of both.

Calcium sulphate (gypsum) is usually recommended at five hundred to one thousand pounds per acre unless extreme problems exist, such as very high sodium. Calcium sulphate can be an excellent product if used properly.

Correctly managed, limestone application can be the backbone of the total soil fertility program. It sets the stage for plant nutrition. Liming may be considered the thermostat in the regulation of the entire soil environment. Here are several general statements that may help clarify the importance of this area of management.

1. Soil pH will normally adjust itself on most soils if we have the opportunity to adjust both calcium and magnesium through application of limestone. This applies where calcium, magnesium and soil pH are all too low. It is almost certain that soil pH will be low if both calcium and magnesium are low because hydrogen moves onto the soil colloid as other cations become depleted. **Even though more emphasis is placed on cation balance than on soil pH we should understand that a low soil pH will cause problems in addition to cation deficiencies.**

Aluminum and iron solubility increases as soil pH lowers. Increased solubility of iron and aluminum can create a toxicity in the growing crop, as well as combining with phosphorous. Iron and aluminum produce complex phosphorous compounds that can become nearly totally unavailable to the growing plant. Broadcast phosphorous quickly becomes unavailable.

2. Low soil pH can release excessive amounts of elements such as manganese, copper and boron. On certain soils these elements can reach toxic levels in availability. **Molybdenum** is the only trace element that becomes less available under a low pH. Legumes in particular can be deficient in molybdenum if the soil pH goes below 6.0.

3. Low soil pH can adversely affect some very important forms of soil organisms. Earthworms cannot tolerate low calcium or low soil pH. The nitrifying bacteria, nitrosomonas and nitrabacters, responsible for converting ammonia to nitrites are adversely affected by low soil pH. The important sulfur-oxidizing organisms, thiobacillus, are also harmed.

The adverse affect on the soil organisms may result from calcium deficiency as well as low pH. Arguing this point could be academic since low pH is usually associated with low calcium. But we do know that most soil organisms such as the nitrifying bacteria require calcium.

4. Calcium and magnesium are the two most abundant cations in the soil pantry with calcium being the larger. We believe that when a soil is properly balanced an ideal ratio for most crops is approximately four parts calcium to one part magnesium.

Because calcium is by far the most dominant cation in the soil pantry, it plays many very important functions within the total soil-plant system. It may be the single most important soil and plant element. As with all nutrients, calcium must be transferred from the soil into the living plant so it can be transmitted to all of the higher forms of consumers. The living soil is dependent upon calcium. Soil organisms, such as those involved in the decomposition of organic matter, nitrification, sulfur oxidation, etc., all demand calcium to function.

Calcium plays many roles within the living cells. It appears to be necessary in serving as a regulator for transporting nutrients across or through cell wall membranes. Calcium, as calcium pectate, must dominate within the middle lamella cell wall or the cellular functions fail because the nutrient transport or transfer system fails. Membranes high in calcium rather than magnesium must be maintained in both plants and animals.

The purging of a digestive tract with magnesium compounds such as epsom salts ($MgSO_4$) is a good example of the results when magnesium displaces calcium in the regulatory membrane. Magnesium membranes cannot regulate. Liquids from the blood seep through the magnesium membranes into the digestive tract, serving as a lubricant to relieve constipation. Visualize, if you will, the same effect on plant cells when soil calcium is low and magnesium is high. This is another reason why limestone should not be applied because of pH alone. It is also a strong point for the selection of the right kind of limestone to fit the soil's needs.

Magnesium plays an extremely important role in chlorophyll production. Each individual chloroplast, or chlorophyll molecule, contains one atom of magnesium. The single chloroplasts represent individual factory units that produce sugars by the process of photosynthesis. Magnesium is the only metal element found in chlorophyll. Without magnesium there could be no photosynthesis and, consequently, no life on earth as we now know it.

5. We are concerned with a low soil pH primarily because it is usually associated with either low calcium or low calcium plus low magnesium, as well as certain adverse affects that a low pH may have on the soil system. Instead of a low pH many soil pH's are high because of excessive applications of limestone or because of naturally occurring excesses of calcium, magnesium or sodium.

Usually a high soil pH would be preferable to a very low pH. However, it is more difficult and expensive to lower a high pH than to raise a low one. A good soil test will identify the cause of either a high or low soil pH. Excessive sodium pushes the soil pH to very high levels and is the most difficult element to manage.

6. The pH number by itself tells nothing about the cation balance in the soil. It would be helpful if a "quality rating" could be assigned to all soil pH numbers.

Let us look at an example of the different cation levels found in a given soil pH. While we believe a soil pH of 6.2 to 6.5 is near ideal for most crops, there will always be exceptions. Figure 122 shows

SOIL pH	% BASE SATURATION Ca	Mg	H
6.3	65%	14%	11%
6.3	43%	32%	11%

Fig. 122

how cations can vary with a 6.3 soil pH. You will notice that there can be a great difference in calcium and magnesium levels for any given soil pH. The pH of 6.3, if it is represented by 65% calcium and 14% magnesium, can be considered a desirable or "quality soil pH." But, a 6.3 pH which consists of a low 40% calcium saturation would represent a low or poor quality pH. Low calcium with high magnesium is a combination we want to avoid.

Magnesium increases soil pH more than calcium because it displaces more hydrogen, therefore, the higher the magnesium, the higher the soil pH will be. If calcium were increased from forty to sixty-five percent in Figure 122 and the magnesium remained at thirty-seven percent the soil pH would probably be over the neutral 7.0. Figure 122 clearly illustrates the fallacy of applying limestone for reasons of soil pH only. Let us build a "quality soil pH" by managing calcium and magnesium.

In general **four** serious mistakes can be made concerning liming. **We can over-lime, under-lime, lime by pH alone,** or **select the wrong type of lime.** From now on, it is my hope that you will never again be guilty of any of these errors.

Living roots can easily overcome high soil pH. Roots must give off hydrogen in order to take on other cations. As a result of hydrogen release the pH will be lowered around the absorbing points on a living root. On all soils efforts should be made to develop large, expansive root systems, and since high pH can reduce trace and other element availability, the development of a large root system is especially important under high soil pH conditions.

Now, with a basic understanding of calcium, magnesium, soil pH and limestone materials, it is time to interpret the soil analysis and make limestone recommendations as needed. Limestone recommendations will consider the soil calcium and magnesium levels, as well as the cation exchange capacity.

Since we want to fill the pantry with a certain percentage of both calcium and magnesium we must remember that the size of the CEC, or pantry, directly relates to the amount of limestone we will need to bring about a soil balance. It will require less pounds of calcium to fill a small CEC than it will take to fill a large one. This is important to remember because excessive liming is a score against good management. Liming must be done right, or not at all, because we usually have enough problems without creating new and unnecessary ones.

First, we must establish a percent base saturation for calcium that relates to the CEC number. Figure 123 outlines a "rule of thumb" for relating the percent base saturation to CEC.

CATION EXCHANGE CAPACITY & CALCIUM BASE SATURATION	
SOIL C.E.C.	% CALCIUM BASE SATURATION
5 OR UNDER C.E.C.	60 TO 65 %
6 TO 10 C.E.C.	65 TO 68 %
11 TO 20 C.E.C.	70 %
OVER 20 C.E.C.	72 TO 75%

Fig. 123

Because we do not really know what an ideal or perfect base saturation should be, some persons may disagree with the figures shown and we might disagree with theirs. These figures are offered to give a general basis for figuring calcium needs and by following these guidelines you should be pleased with the results for most common agricultural and specialty crops including alfalfa.

As stated earlier, soil test readings are given in parts per million (PPM). By multiplying PPM by two you will know how many pounds of calcium per acre are available at present. You must know this figure. We multiply the PPM by two to get the pounds per acre because 1 PPM equals two pounds per acre.

As a sample problem, let us use the following figures:

1. A 10 CEC soil.
2. A soil test showing 715 PPM of calcium.
3. The percent base saturation for calcium is 55%.
4. The magnesium level is high.

Q. How many pounds of calcium should a 10 CEC soil contain?

A. By checking Figure 123 we concluded that a 10 CEC soil requires a calcium base saturation of 68%. If we can figure how much a 10 CEC soil can hold when it is 100% filled with calcium it will be simple to figure how many pounds a 68% fill would be. As shown in Figure 105 it will require 400 pounds of calcium to fill one CEC 100%. Therefore, it will require 4000 pounds of calcium to completely fill a 10 CEC soil (400 × 10 = 4000). A 10 CEC soil at 68% fill needs 2720 pounds of calcium (4000 × 68% = 2720).

Q. How many pounds do you have at present?

A. 715 PPM × 2 = 1430 pounds of calcium per acre.

Q. How many pounds are needed?

A. 2720 pounds minus 1430 equals 1290 pounds deficient or needed. This will raise the present soil level of 1430 pounds (55% base saturation) to 2720 pounds or 68% base saturation.

Q. What kind of limestone should be applied?

A. Since the soils are high in magnesium, a limestone that is high in calcium, but as low as possible in magnesium.

Q. How much limestone should be applied?

A. We learned earlier than pure calcium carbonate contains 40% calcium and that calcium agricultural limestone can be figured at 40% calcium. One ton of calcium limestone would contain 800 pounds of calcium (2000 × 40% = 800). One and one-half tons of calcium limestone is needed (1200 ÷ 800 = 1.5). We rounded the 1290 pounds to 1200.

When using by-products it is important to know the calcium or calcium-carbonate levels for figuring the application rates. The 1.5 tons **assumes** that all of the calcium will become available. For this to happen, the limestone would have to be in powder form or 100 to 200-mesh grind. Therefore, the recommendations would be for 1.5 tons of powdered calcium limestone per acre.

As a general "rule of thumb," if a regular grind agricultural limestone were used, the powdered rate would be doubled and three tons of agricultural-grind calcium limestone per acre would be used. The three tons of regular agricultural limestone will not raise the calcium as quickly as the 1.5 tons of powdered limestone. Because there is probably only about 400 to 500 pounds of powder in a ton of agricultural limestone and the balance ranges from a fine to a very coarse material, it will take longer to react. Many of the larger chips may not break down in a lifetime.

The finer the limestone particles are, the greater amount of surface area per pound exposed to soil acids. The fine particles are important because they react quickly to correct soil deficiencies. Spreading limestone on a windy day may greatly affect results if the finer particles are blown off the field by the wind.

The amount of time it takes a soil to react on the soil test is not so important. After applying the 1.5 (or three) tons, a favorable response will be seen before a complete buildup reflects on a soil test.

Three tons may be the maximum amount that should be applied at one time. In a situation where five tons per acre is needed, three tons could be applied, followed by another two tons, twelve to eighteen months later. Too heavy an application can cause sudden problems, such as phosphorous lockup.

Sudden, massive soil changes should be avoided if possible. Applications of limestone should be done more frequently and at lower rates. We should not let the soil needs exceed a one-ton per acre requirement for regular agricultural grade limestone. Proper maintenance is good business for calcium or any other plant nutrient.

If magnesium levels are also low, magnesium or dolomitic limestone should be used. Dolomitic limestone contains magnesium carbonate as well as calcium carbonate. Dolomite will contain less calcium per ton, depending upon the level of magnesium.

How much magnesium should soils contain? As noted earlier, magnesium should be about twenty-five percent of the correct calcium levels for a given soil CEC. This would give the one to four ratio of magnesium to calcium. This ratio will not always be possible to control but we should try to be as close to it as possible. Remember, the potassium to magnesium ratio is also important.

Calcium needs would be based upon the proper percent base saturation for your particular CEC, as given in Figure 123. The base saturation relates to pounds per acre. The one-to-four ratio relates to pounds, or PPM, of calcium and magnesium, not the percent base saturation. A magnesium base saturation of twelve to sixteen percent is a normal or adequate range for most soils.

We must be practical. Let us suppose that your magnesium levels are a little low and you need three tons of agricultural limestone per acre. In this case, simply apply one ton of dolomite and two tons of calcium limestone. Another soil test in two to three years will tell if you have additional needs. There is much more room for error in the adjustment of calcium and magnesium than there is with potassium.

Magnesium needs are figured in the same way as calcium needs. As shown in Figure 105, it requires only two-hundred forty pounds of magnesium to totally fill one CEC while it takes four-hundred pounds of calcium to do the same.

If a major adjustment of limestone totaling three or more tons per acre must be made it may take three or four years before the full impact of the applications will reflect on a soil test. You must not make the mistake of applying more limestone because a test two years later shows a need. When soil limestone needs are kept within the one ton per acre range you will not have this concern. Under such conditions the one ton application should carry the soils for three to six years, depending upon rotation, use of nitrogen, etc.

In closing, a few words of caution! It is possible that a limestone which is available locally may eventually become the wrong limestone for your farm. If you are using a calcium carbonate limestone, eventually a dolomitic source may have to be located as your soil magnesium levels get lower. A high magnesium limestone may build soil magnesium levels too high, forcing a change to calcium carbonate limestone. Be aware of this possibility.

Some farms have been overlimed with calcium limestone to the point where calcium may exceed eighty percent base saturation with magnesium being low. These farms usually turn to expensive materials such as Sul-po-mag, magnesium oxide or magnesium sulphate to supply their magnesium.

If you have very high calcium and low magnesium levels due to a natural soil condition or improper liming you may want to experiment by applying one ton of high magnesium limestone per acre. You may find, as we have, that the soil magnesium levels will increase while the calcium levels may actually decrease in one or two years. It appears that the magnesium will actually replace calcium on the soil colloids. At the very high levels calcium is held less tightly to the soil colloid and can be replaced easily. It is possible that a one ton rate of dolomitic limestone may be the most economical method of building magnesium.

Be aware that limestone content may be given either as percent of calcium and magnesium or as percent of calcium carbonate and magnesium carbonate. Do not confuse the two in figuring limestone needs.

Our sample problem also points out the importance of using the CEC to determine calcium and magnesium needs. We used a ten CEC soil sample. If we had used a five CEC soil example, the calcium and limestone needs would have been reduced because the pantry would be half as large. This, too, points out why liming by pH alone simply has little basis for logic. For instance, it is possible for a high sodium soil to have an 8.0 or higher pH but be low in calcium. **The 8-plus pH would be no reason to ignore calcium.**

This concludes the study of the upper half of the soil analysis report shown in Figure 96. We will now proceed to the lower portion of the report which shows the soil levels of sulfur and the trace elements, zinc, manganese, iron, copper and boron.

SULFUR

Sulfur is listed on the bottom section of the soil analysis report along with the trace elements. The soil report shows the test results as PPM of sulfur (S). Plants utilize or take in sulfur in the sulphate (SO_4=) form. Elemental sulfur is oxidized into the plant-usable sulphate form by sulfur-oxidizing organisms known as thiobacillus. This important organism functions best when the soil is warm, moist and well aerated.

Since plants utilize sulfur in the sulphate form, we must be able to convert the soil test PPM of S to pounds per acre of sulphates. Sulfur PPM must be multiplied by two to convert the reading to pounds of sulfur per acre. If the soil test shows thirty PPM, there will be sixty pounds of elemental sulfur per acre. Next, we need to know that one pound of sulfur will be oxidized by thiobacillus bacteria into three pounds of sulphates. The thirty PPM on the soil test represents sixty pounds of sulfur or one-hundred eighty pounds of sulphates (30 PPM × 2 = 60 pounds of S/acre × 3 = 180 pounds of sulphates).

Sulfur must be considered a major element along with N, P and K. Many crops, such as soybeans, use more sulphates than P_2O_5.

Sulfur plays many roles in plant nutrition with one of its major roles being its effect on protein formation. There is an intricate relationship between sulfur and its effect on nitrate accumulation, crude protein levels and the biological value of plant proteins. This complex relationship will have a bearing on forage management practices, yields, and nutrient quality.

Sulfur is essential for the conversion of nitrogen into amino acids and the linkage of these amino acids into complete proteins. When sulfur is low the nitrates accumulate in the plant tissue instead of forming into amino acids and protein.

The accumulation of nitrates in forages influences feed quality in two ways. First, high nitrates can cause serious health problems in livestock. They accumulate on the blood hemoglobin and block out oxygen resulting in an oxygen deficiency, or asphyxiation, depending on the degree of hemoglobin saturation. Second, nitrate accumulation can lead to a false crude protein test. Crude protein is a measurement of the total nitrogen in a feed multiplied by 6.25. A high crude protein test in forages may signal a high level of non-protein nitrogen. Non-protein nitrogen in feeds can be called "asphyxiation protein" because of its effect on the oxygen-carrying capacity of hemoglobin.

The sulfur-nitrogen ratio in feeds may be a good means of determining the quality of protein in feeds as well as the presence of non-protein nitrogen. Forages should contain one part of sulfur to ten to twelve parts of nitrogen. This would indicate a good conversion of nitrates into amino acids and protein. The wider the sulfur to nitrogen ratio is, the more likely the feed will contain non-protein nitrogen.

If sulfur is low non-protein nitrogen, as nitrates or nitrites, will accumulate in forages. Nitrate poisoning is a major concern of dairymen and beef producers. It can exist at toxic levels or at lower levels that do not result in toxic symptoms, but reduce production and cause poor health. Adequate soil sulphate levels will virtually eliminate non-protein nitrogen in forages.

Sulfur is not only necessary for the conversion of nitrogen into amino acids, but also is a constituent of the two sulfur-bearing amino acids, cysteine and methionine. These two sulfur-bearing amino acids are essential to all animal life and their presence determines the biological value of a protein. An animal cannot live very long without these two sulfur-bearing amino acids.

Alfalfa contains the amino acids cysteine and methionine. When alfalfa is grown on sulfur-deficient soils it will become deficient in these amino acids and result in a lowering of the "biological value" of its protein. Many plants produce proteins that contain neither of the two essential sulfur-bearing amino acids and these plants have a low biological protein value and must be supplemented when they are fed to animals.

Sulfur is necessary for the formation of chlorophyll and the rate of photosynthesis. Plants with enough sulfur have an exceptionally deep green coloration that can almost be called a sulfur trademark.

Sulfur is part of the defense mechanism of plants. Low sulfur will invite the invasion of insects and disease in crops. It also plays an important role in the formation of carotene in plants. Carotene is converted to vitamin A within an animal's body.

Because of its many vital functions we often refer to sulfur as the "quality element" for crops and consider it to be a most neglected plant nutrient.

In addition to being such a necessary ingredient for many plant functions sulfur is very important to the living soil system because soil organisms are also dependent upon it for their bodies. Many soil organisms could not exist without sulfur to build their body protein.

The amount of sulfur needed to maintain the needs of the soil-plant system is debatable. We know that sulfur and nitrogen are companion elements and some authorities believe the sulfur-to-nitrogen ratio of one to ten or twelve is right for both the soil and plants. Others link sulfur usage to nitrogen rates and may recommend the application of one pound of sulfur for each ten pounds of nitrogen.

In making sulfur recommendations we must consider both the crop and the living soil systems. It is simple to figure crop needs and removals, but nearly impossible to estimate the sulfur needs of all of the living organisms within the soil system. Populations of soil organisms are regulated by several environmental factors including nutrients such as carbon, nitrogen, sulfur, etc. One wonders what limitations sulfur could be placing on these systems, and consequently, on both crop yields and quality.

Sulfur deficiencies may occur due to several situations. Under cool, wet, or poorly-aerated conditions, soil sulfur will not convert to the plant-usable sulphate form. Anerobic soil conditions will reduce the sulphates to sulfur gasses. These gasses are not only unavailable to plants, but also subject to gasification losses from the soil. Sulphates may leach easily, particularly on light-textured, low-organic soils. These are some of the more common conditions or circumstances that can lead to a deficiency of sulfur.

The key to sulfur management is defining the needs and then the selection of the product to best satisfy those needs. Sulfur needs are defined with soil tests and tissue analysis. While the interpretations of need and sulfur recommendations are demanding, it is difficult to apply too much sulfur. Under certain soil conditions, such as high sodium, more than five-hundred pounds per acre may be recommended with no adverse effects on the growing crop.

As with soil pH, sulfur levels will fluctuate considerably during the season. This fluctuation is caused by many factors and will vary greatly, depending on conditions. Soil life and the growing crop's demands for sulfur will lower the soil levels. We think soil tests will more accurately represent true soil levels when taken during the off-seasons of late fall or early spring. This assumption could be debated, but has been concluded from many years of soil auditing and observations of crop responses.

We have established a soil reading of thirty to thirty-five PPM of S as desirable. This level does not necessarily relate to the standards of soil testing laboratories.

Since most crops need at least as much if not more sulphates as they need of phosphates, and considering the soil microbiological needs, we believe **the soil sulphate level should be at least as high as an acceptable P_2O_5 reading on the P2 test.**

At a desirable P2 phosphorous level of fifty PPM, we would have two-hundred thirty pounds of P_2O_5 per acre (50 PPM $\times$ 4.6 = 230). At thirty-five PPM on the sulfur test, we would have two-hundred ten pounds of sulphates per acre (35 PPM $\times$ 2 = 70 pounds S per acre $\times$ 3 = 210 pounds sulphates). Since total soil and plant demands for sulphates appear to be as high or higher than for phosphorous these sulphate levels seem practical.

Once soil levels have been built, it will require about twenty pounds of elemental sulfur per acre each year to maintain losses from crops, leaching, gasification, erosion, etc. This can vary of course, depending on crops and yields. A ten ton alfalfa yield, for example, would remove at least sixty pounds of sulfur per acre. Many farmers apply no sulfur in their alfalfa management programs which in view of alfalfa needs can only be regarded as poor management.

The acidulation, or pH lowering, effect of twenty or more pounds of sulfur per acre is minimal. As long as proper calcium and magnesium levels are maintained we need not be concerned with the pH factor.

Sulfur management must include the determination of the soil buildup needed, building the proper levels and then maintenance to replace losses on an annual or bi-annual basis.

What is the best method of building and maintenance of soil sulfur? We must first determine the original application necessary to build a sulfur base that is sufficient to meet the demands of the soil-plant system.

Let us assume that we want a soil test level of thirty-five PPM of sulfur and the soil test reads twenty PPM. A buildup of fifteen PPM or thirty pounds of elemental sulfur per acre is needed. The thirty pounds of sulfur can be applied either in the elemental or in the sulphate form. Elemental sulfur is available in liquid, powder, or granular forms for general soil application.

Sulfur is available in liquid and flowable forms that can be blended with liquid nitrogen or mixed with N, P and K liquid fertilizers. Flowable sulfurs are available in analysis exceeding fifty percent sulfur. There are several ammoniated sulfur solutions, such as ammonium thiosulphate, which contains twelve percent nitrogen and twenty-six percent sulfur. These materials can be broadcast or mixed into liquid starters and banded alongside the row.

Liquid sulfurs are convenient for handling, but their cost per pound of sulfur is usually considerably higher than the dry forms of sulfur. Liquid sulfurs contain extremely small particle sizes and should convert quickly to sulphate, providing soil conditions are favorable.

Dry forms of sulfur are available as wettable powder or granules. Their sulfur content ranges from ninety to ninety-five percent. Wettable powders can be applied as suspensions, or even blended into dry broadcast fertilizers along with an antidusting additive. Fertilizer grade granules can be blended into broadcast or applied at the rate of ten to twenty pounds per acre in starter fertilizers. The quality of the granules is important because the granules must dissolve to expose the fine sulfur particles to the action of soil-oxidizing organisms. Particle size is as important for sulfur as it is for limestone. The size of the particle determines the surface exposure per unit of material, with the finer particles converting more quickly to plant-usable forms.

Among the more popular forms of sulphated sulfur are sul-po-mag, sulphate of potash (0-0-50), calcium sulphate and ammonium sulphate (21-0-0-24). With the exception of ammonium sulphate, all of these materials were discussed in this unit. These products are normally used to supply magnesium, potassium or calcium, with the sulphate being a fringe benefit. They fit only when the other elements are needed.

Ammonium sulphate (NH_4SO_4) is an excellent source of sulphate with broad application potential since most crops need supplemental nitrogen. **Ammonium sulphate provides immediate availability of sulphate along with nitrogen, making it an excellent product for use on forage crops such as alfalfa or pastures.** Adding nitrogen with elemental sulfur to forage crops could cause nitrate or non-protein nitrogen buildup because the sulfur would require **four to six weeks** to convert to the plant-usable sulphate form.

Sulfur is extremely important for all legumes. The rhizobia, or nitrogen-fixing bacteria, within the root nodules on legumes must have an adequate supply of sulfur to fix nitrogen from the air. Most legumes do not fix enough nitrogen for maximum yields. Additional nitrogen for legumes such as beans and alfalfa can be very profitable only if there is adequate sulfur available.

Ammonium sulphate contains twenty-one percent nitrogen along with twenty-four percent sulfur in the ready-to-use sulphate form. It is an ideal source of sulfur that can be blended and broadcast with other fertilizers, such as 0-0-60.

Many times we have observed grasses crowding out legumes such as alfalfa. This is often blamed on low potassium since at low soil levels grasses recover potassium more easily than alfalfa. This same principle applies to sulphates. Grasses take up sulphates more easily than alfalfa. Low sulfur has caused many alfalfa fields to turn to grass within three years. Once the value of the nitrogen is subtracted, one may find that ammonium sulphate provides a very low cost source of sulfur.

How much sulfur should be applied for maintenance once soil levels have been built? We believe that normal crop systems on most soils will require about twenty pounds of sulfur per acre each year to maintain soil levels. An application of **one-hundred fifty pounds** of ammonium sulphate or its equivalent, every other year, should maintain soils very well. On a corn and bean rotation the application of one-hundred fifty pounds of 21-0-0 is recommended **ahead** of the soybeans because of their high needs of both sulfur and nitrogen.

On medium to heavy-textured soils we prefer to apply elemental sulfur in the fall because heavier soils resist leaching and fall application allows more time for sulfur to convert to the plant-usable sulphates. Spring applications of all fertilizers, including sulfur, is preferred for light-textured soils to minimize losses by leaching.

The form of sulfur used makes little difference, providing it becomes available to the plant to **complement nitrogen uptake.** This means that elemental sulfur must be converted to sulphate by sulfur-oxidizing soil bacteria. It is difficult to predict the sulfur oxidation potential on a given soil. For this reason we prefer the sulphate forms, particularly during the soil sulfur-building phase.

Foliar-applied sulfur has fungicidal properties, as well as providing sulfur nutrition to the plant. The flowable sulfurs are excellent for foliar application.

Corn may turn yellow from low nitrogen, low sulfur, or a combination of both. If a tissue analysis shows low nitrogen and marginal sulfur a foliar application of two quarts of flowable sulfur along with six pounds of liquid urea nitrogen per acre would be beneficial. This combination can work wonders on corn to help it through adverse soil and weather conditions that cause the nitrogen and sulfur shortages. The nitrogen must be in the urea form which is absorbed into the plant tissue within a very few hours. This provides an excellent salvage method and will be discussed further under **Nitrogen Management.**

From now on, let us consider sulfur a major element, along with nitrogen, phosphorous and potassium. Give sulfur management high priority and treat yourself to better yields and quality.

TRACE ELEMENTS

The lower part of the soil test, as shown in Figure 96 on Page 145, relates to the trace elements. We are mainly concerned with the trace element levels of zinc, manganese, iron, copper and boron.

These elements are called trace or minor elements because their levels within the soil and plants are measured in very minute quantities. But we must never make the error of judging their value in proportion to the extent in which they are found within the soil and plants.

Trace elements appear to be primarily catalytic in their functions. They are found within the molecular structure of enzymes and therefore may be involved in virtually every cellular function. They are essential for both plants and animals and may very well be one of the most neglected aspects in both plant and animal nutrition.

During the decade of the 1970's we have observed a rather alarming decline in soil trace element levels throughout the midwestern states. During the same period tissue analysis shows trace element levels in the lower rather than the upper sufficiency range.

When reviewing soil test data from 1960 through 1965 it was not unusual to have 2 PPM on the boron test. By 1980 we felt fortunate if soil tests were showing 1 PPM of boron. Most soil tests taken in the midwest from farms on a continuous corn, or a corn and bean, rotation are testing under 1 PPM for boron. We are finding the same trend on zinc and copper tests.

In our quest for both quantity and quality production we must be aware of the need for total management of all the production factors. Due to the complexities and interaction between the many factors it is nearly impossible to lay credit or blame on any single one. Trace elements are but single factors within the whole. However, in many situations they are neglected factors. The purpose of this section is to provide guidelines for determination and correction of trace element deficiencies.

Soil test data is a necessary first step to determine whether or not a trace element problem exists. Then trace element management must include the proper interpretation of the soil test data to determine needs, the selection of proper trace element forms to fit those needs and, finally, an accurate record keeping system to record the rates and dates of applications.

Naturally occurring trace elements are distributed very uniformly throughout the soil. Broadcast trace elements will not be uniformly distributed. For this reason it may require three or more years before they will become sufficiently distributed to reflect accurately on a soil test. Recording rates and dates of applications becomes essential.

When a given amount of a trace element is correctly applied to bring the soil into balance we can rest assured that the deficiency has been corrected regardless of what the next soil test may indicate. Once the buildup stage has been reached we can turn our attention to soil maintenance. Records are important in the maintenance program also because the rates and frequency of application will vary for each trace element.

We now turn our attention to the detailed management of zinc, manganese, iron, copper and boron. It must be pointed out that the soil levels we recommend for each trace element may not be in full agreement with standards set by others. Due to the differences in crop variety needs and the many problems that limit soil aeration, root extension and nutrient recovery, we may question the validity of any set standards. Because of favorable results over the years, however, we believe that the slightly higher soil level standards that we are recommending are justified.

ZINC

We have set the desired soil levels for zinc at 6 PPM to 8 PPM. Six PPM will normally be adequate and is a good minimum level for which to strive. The soil test reading indicates the PPM of elemental zinc per acre. Since 1 PPM will represent two pounds of elemental zinc per acre we must add two pounds of elemental zinc per acre to raise the soil level by 1 PPM.

Basically there are three methods for satisfying zinc requirements. We may broadcast zinc, add it to starter bands, or apply it as a foliar feed. The method selected will be determined by several factors such as cost, convenience, land tenure, etc.

Whenever possible and when soil conditions permit broadcasting seems to be the best method to build soil levels and supply plant needs. Building soil uniformity and balance along with supportive practices that assure recovery is the best approach to crop production. However, certain soil conditions such as poor drainage, poor aeration and high pH factors, may favor banding, foliar feeding, or a combination of methods.

As we consider broadcasting we will use for an example a soil test showing a reading of 3 PPM for zinc. To build the soil to the 6 PPM level we must apply enough elemental zinc to increase the soil level by 3 PPM. It will require six pounds of elemental zinc to raise the soil levels 3 PPM. Next, the form of zinc and method of application must be decided.

For broadcasting we have the option of using the dry forms of zinc sulphate ($ZnSO_4$) or zinc oxide (ZnO). Both forms are available, but zinc sulphate is by far our first choice. Zinc sulphate is water-soluble and readily ionizes, making the zinc highly available. Zinc oxide is not water-soluble and its availability is very slow and unpredictable.

How much zinc sulphate will be needed to supply the six pounds of elemental zinc required to raise the soil level 3 PPM? Zinc sulphate contains 36% elemental zinc. An easy way to figure it is to apply **six pounds** of zinc sulphate per acre to raise soil levels by **1 PPM.** This will supply a little over 2 lbs. of elemental zinc.

Zinc sulphate may be applied in several different ways. A granular form is made that can be blended into broadcast fertilizers such as 0-0-60, or it may also be purchased in a fine powder form that will dissolve in water or liquid nitrogen. Application with a liquid gives superior distribution.

NOTE: **Never** add a sulphated product such as zinc sulphate to a liquid phosphate. You will end up with a gravel-like precipitate.

Another method of supplying zinc is to add it to the starter fertilizer. Banding concentrates a material, requiring less to do the job. For this reason we would not band at the same rate that we would broadcast since to do so would very likely result in zinc toxicity to the plant.

Obviously, the form of zinc material used will depend on whether a liquid or dry starter is used. For use with a dry starter we would recommend using the zinc sulphate granular form instead of the zinc oxide. The oxide form would not become available until about the same time as harvest, which would be too late. Zinc sulphate granules are designed to blend into dry starters.

For row crops such as corn, beans, milo and sunflowers, soil levels can be built by applying zinc sulphate granules through the starter over a period of two to five years. In our example, where a total of eighteen pounds was needed, four to five pounds of zinc sulphate granules per acre could be blended into the dry starter for a period of three to four years, or until a total of eighteen pounds per acre had been applied.

The four to five pounds of zinc sulphate will take care of the crop needs on a year-to-year basis while permitting the cost to be spread over three to four years. This is an excellent approach when operating on a year-to-year crop lease. Zinc requirements can be assured at a minimum cost.

It is satisfactory to build soil zinc levels through banding when using conventional tillage. This principle, however, does not apply to no-till crops. This is discussed in the unit on **Fertility Systems.**

After soil levels have been built they must be maintained. Our maintenance guide is very simple. Just apply five pounds of zinc sulphate per acre either broadcast or banded, with conventional tillage, every **fifth** year. Whether zinc is managed by broadcast or banding it is necessary to keep accurate records of the rates and dates of application.

Soil levels can be built more economically with dry starters than with liquid. Liquid zinc is available in chelated forms or in combination with nitrogen. Several companies produce liquid zinc chelates of various analysis. A typical zinc chelate may contain up to one pound of elemental zinc per gallon at a cost of about six dollars per gallon (1981 prices). In comparison, a pound of zinc from zinc sulphate may cost less than seventy-five cents. Obviously, we cannot afford to build soil levels with expensive chelates. However, the chelates can be used efficiently in liquid starters to supply annual zinc needs in combination with soil building or when soil building is not the goal. **One to two quarts** of an 8% to 10% zinc chelate per acre, banded in starter fertilzer, will manage zinc needs on an annual basis for most row crops where soil deficiencies are not too severe.

Another method of supplying zinc is through foliar application. Foliar application using liquid forms especially designed for this type of application is an excellent method for use on orchards and vegetable crops. It is difficult to get certain nutrients into the root zone on established orchards, leaving little choice except foliar applications. Many specialty crops have very limited root systems making foliar applications effective.

For general farm crops trace elements should come via the soil route. Foliar applications under certain conditions may be necessary as a salvage tool, but should not be generally considered a method of supplying the crop needs.

MANGANESE

Manganese is essential for chlorophyll and photosynthesis. Its availability to the plant is greatly reduced by high soil pH or cold soil conditions. When soil levels are marginal visual signs of deficiencies may appear in the crop early in the season if weather conditions are unfavorably cool or wet. Certain crops such as beans, as shown in Figure 124, will show the yellowing effect of manganese deficiency more quickly than crops such as corn. It must be remembered, however, that a crop can be deficient even though it is not detected with the eyes. These are called "hidden deficiencies."

Fig. 124

We have set the soil test level for manganese at a minimum of 25 PPM and would prefer it nearer to 50 PPM. At the lower range we should be aware of possible plant deficiencies early in the season when the soil may be cool and root systems small.

You will note that manganese soil levels are several times higher than zinc, copper or boron. Because of this it may be too expensive to try to build a soil level that is low. It would require forty pounds of elemental manganese to raise a soil level from 10 PPM to 30 PPM. Using low cost manganese sulphate, at 1981 prices, this would cost about forty dollars per acre. This limits our application options to either banding or foliar feeding.

Manganese is available in either dry or liquid forms. Dry manganese, as manganese sulphate, contains 28% elemental manganese and is available as a granule or powder. Manganese sulphate is available in a powdered form known as tecmangam that can be used as a foliar. Tecmangam can be dusted or dissolved in water and sprayed. A normal foliar application for most crops would be 10 to 15 pounds of material per acre.

Manganese sulphate can also be found in various granule sizes. There is a relatively new granular form called "Ammoniator's Special Manganese Sulphate" designed to be applied through granular insecticide applicators. A ten to fifteen-pound application can be dropped in a narrow band directly over the row while planting. This method is most effective on light-textured soils. Other granules are available for blending into dry starter fertilizers at ten to twenty pounds of material per acre.

Manganese may also be obtained in liquid chelate forms containing five percent or more of elemental manganese. The chelates are designed primarily for foliar feeding and are quite effective at one to two quarts per acre.

The need for supplemental manganese usually occurs early in the season, particularly under cool, wet conditions. Manganese deficient plants will lose their chlorophyll (chlorosis) and turn various shades of yellow. Once the soil warms and roots begin to develop the symptoms will normally disappear. Manganese deficient plants must be salvaged by foliar feeding. Once deficiency signs appear we can assume there will be a loss of yield. Prevention is the key. If we know soils are in the low range preventative steps should be taken to head off both visual and hidden deficiencies.

Increasing soil pH reduces manganese availability. For this reason applications of too much limestone can be particularly damaging when the soils are marginal in manganese. Soils with a 6.5 or higher pH will show manganese deficiency much sooner than soils with a 6.0 pH.

Again, we must keep accurate records as to the rates and dates of application. If ten pounds of 28% manganese sulphate per acre has been applied it will amount to 2.8 pounds of elemental manganese. Ten years of applications at this rate will amount to twenty-eight pounds of manganese which will increase soil levels by fourteen PPM.

COPPER

Copper has been almost totally ignored in general agriculture, yet it plays an extremely important role during the plant's reproductive stages and in photosynthesis. In the past copper served as a fungicide and was used extensively in orchards and on specialty crops. With this practice excesses were more common than deficiencies.

We have set the soil level standards for copper between 2 PPM and 3 PPM. We believe 2 PPM should be the minimum and prefer to work toward the 3 PPM level. Soil test levels throughout the midwest are generally showing less than 2 PPM on the copper tests. Tissue tests for soybeans are often reflecting the lower soil levels. The same trend is seen in corn and small grains.

The crystal form may be blended into dry fertilizers such as 0-0-60 and broadcast. This is satisfactory providing a crystal size can be selected that will spread uniformly with the granular fertilizer carrier. The powdered form along with the correct binder-sticker material may also be blended and broadcast with other dry granular fertilizers.

Copper sulphate contains twenty-five percent elemental copper and is available in granule, crystal and powder forms. The powder form can be dissolved in water or liquid nitrogen and applied in a weed-and-feed program. This makes an excellent method of application because of the uniformity of application a spray system can offer.

Copper is available in liquid or dry forms and can be broadcast, used in starters, or applied as a foliar feed. Both starters and foliars have limitations as a means of supplying copper to general farm crops. Our recommendation would be to build the soil levels with copper sulphate.

Since copper sulphate contains twenty-five percent of elemental copper it will require four pounds to provide one pound of element. The one pound of element would raise the soil level by one-half PPM.

Once the soils have been built to the desired level of two to three PPM we recommend that they be maintained by applying one pound of elemental copper every fifth year.

Copper sulphate crystals can be blended into a dry starter, but it is preferable that they be broadcast. It is too easy to reach toxic levels in a band. Too much copper in a starter band could block manganese and cause early chlorosis.

Liquid copper chelates may be used in liquid starters at one to two quarts per acre. Liquid chelates are low analysis and more expensive. The chelates are used extensively in orchards and specialty crop areas. We would not recommend the broadcast application of chelates as a means of building soil levels.

BORON

Boron plays several key roles within the plant, including pollen viability, good seed set, normal terminal growth, photosynthesis and movement of sugars. It is one of the most neglected of the trace elements because too often its importance is related only to tuber crops or alfalfa. It is a serious mistake to believe that boron is not needed for common farm crops such as corn or soybeans.

Boron levels are also becoming quite low in many midwestern states. We believe the soil test should show between 2.5 PPM and 3 PPM for general crops. Three PPM represents six pounds of elemental boron per acre.

Boron can be very toxic and this seems to be particularly true in the presence of low calcium. Certain crops are much more sensitive to boron toxicity than others.

The application of boron has more limitations than the other trace elements. In general it is preferred to limit its application to broadcast even though it can be used in starter bands or foliar. Boron is available mainly in dry forms, but there are now some excellent liquid or flowable forms available. Stoller Chemical Company, for example, provides liquid flowable boron either as straight boron or in combination with liquid calcium. The boron-calcium combination is best for foliar application.

Dry boron comes in granules or powdered forms. Granules are generally available in the twenty-one percent or twelve percent boron concentrations. The 12.5 percent material is most popular for broadcasting. The lower concentration offers more volume for blending, resulting in a more uniform spreading pattern.

Sol-u-bor is a twenty-one percent powder that is designed to be broadcast in water or liquid fertilizers. It must be suspended with agitation in the tank. Sol-u-bor is an excellent product because it can be applied very uniformly. Uniformity is important when working with such small quantities.

Sol-u-bor is also used as a foliar source of boron in crops like potatoes. It can be applied alone or with fungicides at the rate of one-half pound of material per acre. At this rate it can be applied two or three times during the growing season as needed.

Certain crops, such as beans, are more sensitive to boron toxicity than others. Alfalfa and root crops require high boron and show less sensitivity. Almost all crops are most sensitive to boron toxicity during germination and seedling stages of growth. Because of this sensitivity during early growth it is preferable not to add boron to starter bands. While boron, at less than one-half pound of element per acre, can be effective in starter bands for corn and other less sensitive crops, the margin for error is very narrow. Under most conditions boron should be broadcast to build the soil levels.

The amount of boron applied at any one time will depend upon the present soil levels, crop to be grown and the time of year it is to be applied. The following guidelines are offered for the application of boron.

Broadcast two pounds of elemental boron per acre during the fall or one pound per acre if applying in the spring. The normal application would be fifteen to twenty pounds of 12.5% boron material per acre if applied in the fall or ten pounds of the same material if spring applied. These one or two pound rates are general guidelines only. The ten pounds of 12.5% material would supply 1.25 pounds of element per acre. This is slightly over the recommended spring rate of one pound per acre but presents no problem for corn or soybeans. The spring application is assumed to be ahead of spring tillage and incorporated at least five inches in depth.

If the twenty-one percent sol-u-bor powder or granules are used it will take five pounds of material to get one pound of elemental boron. To increase soil levels by 1 PPM it is necessary to apply two pounds of elemental boron per acre. This would require one two-pound fall application or two one-pound per acre spring applications over two seasons.

After the soil levels have been built up we would recommend the application of one pound of elemental boron per acre every five years on medium to heavy-textured soils. One pound of elemental boron per acre every three years would be the guideline for light-textured or low-organic soils.

The importance of record-keeping cannot be overstressed. Without records as to the dates and the rates of applications, maintenance could be neglected. This information should be transferred to new land owners or operators.

We should caution that a low soil pH increases boron solubility or availability. If the low soil pH is accompanied by low calcium the potential for boron toxicity increases. Limestone needs should probably be supplied before applying boron.

IRON

Iron is an essential element in building the chlorophyll molecule. Without a single molecule of magnesium there can be no unit of chlorophyll (chloroplast). Iron appears to be necessary to fixing magnesium into the chloroplast.

A lack of iron causes a yellowing or iron chlorosis, due to the lack of chlorophyll. Without chlorophyll the plant cannot produce sugar and without sugar from the "factory," the plant cannot survive. Iron deficiency, or iron chlorosis, occurs mostly on calcaeous, alkaline soils with a pH of 7.5 or above. These soils may be high in sodium or extremely high in potassium, calcium and/or magnesium. As with manganese, iron deficiency is most likely to occur during very early growth under cold or wet soil conditions. Excessive phosphate applications could also trigger iron deficiency in acid soils.

Certain areas of the country, such as parts of Minnesota and northern Iowa, are more prone to iron deficiency symptoms than others. In these areas farmers are as aware of the iron problems as the farmers in areas where manganese is deficient. They recognize the problem and are prepared to deal with it.

Iron management, as with manganese management, is greatly influenced by soil temperature and root development. A large root system is the best insurance against deficiency. Soil temperatures also directly relate to root growth making iron management as well as manganese management absolutely necessary during the critical early growth period on certain soils.

Iron management is most effective as a foliar application because this assures instant utilization. Iron sulphate or other compounds applied to dry starter bands may not be utilized when they are needed due to cold or wet soil conditions. Foliar application skirts these problems by being applied directly onto the leaf. By knowing the conditions that cause an iron deficiency we can be prepared to take action before the deficiency symptoms become too pronounced. Preventive maintenance is the best medicine and everything possible should be done to remove the cause of any deficiency. Anything done to improve drainage, soil aeration and soil temperature will minimize or even eliminate the problem. Ridge planting can be a very effective system under adverse soil conditions.

Iron chelates are available and very effective as foliar sprays to correct iron deficiencies. Iron sulphate can also be dissolved in water and used as a foliar, but it is more difficult to handle and can burn foliage if it is too concentrated. The liquid chelates, used as recommended on the labels, are probably the best remedy.

Applications of trace elements to the soil does not necessarily guarantee recovery in sufficient quantities to satisfy plant needs. Total soil management is necessary.

Since trace elements are used in such small quantities and the soil levels are measured in PPM everything possible must be done for the development of large root systems. Adequate soil oxygen, as stressed earlier, is necessary for the development of massive roots and to assure nutrient uptake because both of these are oxygen-demanding processes. Large root systems are additionally important in high pH soils to produce enough root acids to overcome the high pH effect on nutrient availability.

This concludes our study on soil chemistry as it relates to the soil analysis. The primary purpose of the soil analysis is to define soil chemical problems. Soil testing and the proper interpretation of the test data are the first steps in the process of solving these problems.

From the soil test data we have learned how to determine soil needs. We have also learned that there are different forms of nutrients and different methods of their application. This knowledge can now be put to use in the development of a complete fertility program for our individual farms.

DEVELOPING A FERTILITY PROGRAM

Our goal in managing soil fertility is to supply adequate plant nutrition at a minimum cost. Since the method of application can greatly influence both the fertility rates and end results it becomes very important to understand the concepts of "fertility placement."

Fertility placement relates to "nutrient concentration" and its relationship to soil reactions, soil aeration, root patterns, forms of nutrient and moisture supply. The ultimate objective is to assure adequate nutrient recovery by the growing plant.

For instance, fertility practices should not band nutrients in wheel tracks where a lack of oxygen prevents root growth. It is unprofitable to broadcast a nutrient such as phosphorous where soil conditions will render it unavailable. We must continually remind ourselves that the fertility of our soils must be judged by the crop's response to that fertility. The plant must gather in whatever the farmer and the soil "serves up."

The diagram on Page 143 shows how the root mass can influence the fertility of a soil. Clods altering the soil fertility by locking up nutrients are shown in the bottom figure.

Surface clods, soil aeration, root mass, soil water supply, soil type, competition from decay, nutrient concentration and soil temperature are some of the factors that can influence soil nutrient accessibility and recovery. Good fertility placement is a means of alleviating many of the factors which hinder nutrient recovery. An overall fertility program must be designed to assure recovery.

Because crop production involves a total "systems approach" every input must be part of a well-planned system. A selection is made of specific tools to make up our "tillage system." In a similar manner the overall "fertility program" is also comprised of a number of carefully orchestrated individual systems.

The fertility program includes such entities as "the nitrogen management system," the broadcast system, the start system, a foliar-feeding system, a maintenance system and even a salvage system. Ironically, even the best-laid plans can be shot down by Mother Nature, necessitating salvage attempts.

It is difficult to explain precisely how to lay out a good fertility program because of the many crops and the overlapping of the various fertility systems that make up the whole. All systems are important and deciding where to start is much like trying to decide which came first, the chicken or the egg!

To begin, we can dilute the confusion by thinking strictly in terms of proper placement of nutrients. Next, we can think of placement as either broadcasting or concentrating in bands. This will narrow our application methods to two options — banding or broadcasting.

Some products are better broadcast and others can be best managed by banding. Because of financial considerations and the need to budget we may band certain nutrients that we would normally prefer to broadcast. We must have several options to fit almost any situation. The following points of review will help set the tone for planning your complete fertility program.

1. We prefer to broadcast potassium fertilizers to minimize the salt concentration near the seed or root mass. We want to build high soil levels of potassium coupled with good aeration to assure root growth and recovery. You may feel strongly in favor of using some potash in starter bands and to this we have no objection. The broadcast preference is often bent by banding as much as seventy-five to one-hundred pounds per acre of 0-0-60 or its equivalent. We do not hesitate to add potash to the starter when we know soil aeration will be limited.

For instance, if you are renting ground on a year-to-year lease that is low in potash you have no choice but to band. We have one general rule to follow in banding potash. When fifty pounds per acre of 0-0-60 is exceeded on row crops (thirty-inch or wider spacing), we recommend placing the band three to four inches to the side of the row. This gives the chlorides time to rinse out and become diluted before the roots wander into the zone. A high salt concentration near the roots can cause damage, commonly called "fertilizer burn."

Fertilizer burn is caused by plasmolysis, the opposite of osmosis. The high salts cause nutrients and water to seep out of the roots into the surrounding soil. This can cause plant damage ranging from slight retardation of growth to complete dehydration and the resulting death of the roots and plant.

2. Trace elements can be managed either through starter bands or broadcast, with some exceptions. It may be too expensive to build low soil manganese levels. Boron and copper are best broadcast because the margin for error is small in preventing a toxic condition near the seed. You cannot safely supply enough boron in a starter band to satisfy the needs of most crops if the soil levels are low. Too much copper in the starter band can hinder manganese uptake and result in chlorotic plants. It can also interfere with iron.

3. In our "start system" we recommend certain trace elements even though the soil levels are considered adequate. This is explained in the trace element section relating to the soil analysis. Trace elements are applied in very small amounts and even though enough of an element has been applied to adequately build the soil level it may not reflect on the soil test for two to three years. Keeping

good records as to the rates and dates of these applications cannot be overemphasized because managing maintenance of trace elements demands it.

4. Placement has a definite bearing on nutrient recovery by the plant. A topdress of one-hundred pounds of some fertilizers on such crops as small grains, pasture or hay may be as effective in crop response as three-hundred pounds or more that is incorporated throughout the top seven to eight inches of soil. The dilution factor is enormous if we imagine mixing a relatively small amount of fertilizer with two-million pounds of soil. At best, a root system could cover less than one percent of the soil volume. Banding underground is even more efficient for row crops than topdressing is for sod crops. Banding is more efficient because it is already underground and more likely to remain in a moist condition.

5. Nitrogen is the growth element. An excess of nitrogen accelerates vegetative growth. Excessive top growth can create deficiencies in other nutrients setting the stage for diseases, insects and reduced yields. It is particularly damaging when applied too early in the life of a plant. A small root system cannot gather in a sufficient quantity of nutrients to support a nitrogen-accelerated top growth. Our goal is to grow roots first and avoid forcing top growth until soil conditions favor microbiological activity and nutrient release. Timing and placement of nitrogen are very important. Without question, nitrogen is the most mis-managed element in agriculture.

6. Moisture is essential for root function and nutrient recovery. Most of a plant's needs are removed from the water film surrounding the soil colloid. Once this film reduces to the permanent wilting point the nutrients are no longer available. Most starter fertilizer bands are placed too shallow for maximum recovery of nutrients.

7. The start, or early growth period, for a plant is as important as it is for an animal. Genetic expression is set early in the life of many crops. Approximately five weeks after emergence, corn for example, has already determined the number of rows of corn there will be on the cob.

Everything possible must be done to prevent setbacks or loss of growing time early in the life of our crops. If the vegetative stage is properly grown we can expect better results in the reproductive stage. The importance of nutrient balance at the start must be recognized. Excessive nitrogen can induce deficiencies by forcing top growth at the expense of root growth.

Adverse soil conditions can also affect early nutrient release and availability. For these reasons everything possible to promote a healthy start and cut losses during the first stage of production must be done. During the early growth cycle we have the best opportunity to increase potential yield by removing and minimizing as many yield-limiting factors as possible.

Management of the following areas will complete the process of putting together a total fertility program.

1. Limestone management
2. Broadcast program
3. Seedbed management
4. Seed selection
5. Row support
6. Starter fertilizer and nitrogen

LIMESTONE MANAGEMENT

Liming is normally a separate phase of the soil fertility program and was outlined in detail earlier in this unit. However, we will emphasize again that liming practices must be based primarily on the soil's needs for the cations, calcium and magnesium. Limestone also removes the acid element, hydrogen, therefore raising the soil pH. In review we must think in terms of building a "quality pH" which results in the proper balancing of all of the soil cations, such as calcium, magnesium and potassium.

Liming must never be done strictly for the sole purpose of managing the soil's pH. Limestone which is available locally may eventually become the wrong lime for use on an individual farm. A source of either dolomitic or calcitic limestone may be needed to maintain the best calcium and magnesium ratios. Limestone may be considered a fertilizer because it supplies the plant nutrients, calcium and magnesium, therefore becoming an important part of our fertility management program.

BROADCAST PROGRAM

General fertility goals are concerned with both the building and the maintenance of soil nutrient levels. We usually prefer to broadcast potash to minimize its use in starter fertilizers. We also prefer to build most of the trace elements through broadcast rather than banding. We want to build soil phosphorous levels to 45 PPM or higher while maintaining crop removals through the planter application systems.

There are many exceptions and no hard-fast rules regarding the broadcast program. We will show options as fertility management systems are outlined.

SEEDBED MANAGEMENT

The objectives for preparation of a proper seedbed are outlined in the unit on **Tillage.** Many poor-producing crops got their start in a poorly-prepared seedbed. Seedbed tillage is the final operation which sets the stage for a successful soil-seed marriage and culminates in germination and emergence. Our total efforts are dependent upon getting a sufficient stand at the start.

SEED SELECTION

Selection of seed should command more attention. The problem is not so much a concern for genetic potential as it is for seed vitality. Once the genetic lines have been established the goals of seed production must be to produce physiologically mature seed with a complete nutrient package.

Select large-sized seeds whenever you have a choice. A small seed may, or may not, have the same chemical composition as a large seed which came from the same head or ear. Corn is an example of seeds that vary in size simply because of their genetic position on the cob. The small, tip kernels on an ear of corn will have the same genetics as the large, butt kernels. They may also have the same chemistry and produce equally vigorous seedlings that produce just as well at harvest. However, the butt end kernels are the first to fill and the tip end is the last area of the ear to fill. Progressive stress throughout the season can weaken the plant to the point where it may alter the nutrient balance in those last-to-fill kernels.

Just because the ovule pollinated and the seed formed is no guarantee that it will completely fill. All multiple seed crops, such as corn or wheat, have a filling sequence. Nature tries to completely fill as many seed sites as possible. If all seeds on a head of wheat or ear of corn filled equally and at the same time it could be hazardous for the survival of the species. Mother Nature is not that foolish; she will completely fill a few sites to guarantee survival and do her best on the others. A few seeds may not secure the future of a farmer, but it will guarantee a chance for survival of the species.

A nutrient weak seed will produce a weaker seedling. As in the animal kingdom, the weak cannot compete with the strong and usually will not reproduce. Near harvest time a corn field can be entered and all of the plants located that got off to a slow start.

A slow start can be caused by several factors, such as insect damage, micro-differences in seedbed quality, weak seed, etc. Once an individual plant gets behind it will seldom catch up or produce as well as the strong seedling which may be next to it. In corn these plants show up near harvest by being six or more inches shorter, smaller in diameter, and possibly even totally barren of grain, or a nubbin at best. As with animals, a "runt" just does not make it.

In summary, we conclude with the following thoughts in relation to selection of seed:

1. The size of seed coming from the same plant or variety has no bearing on genetic potential. Genetics govern yield potential but are only one factor among many that determine how well that yield potential will be expressed.

2. The extra nutrients within a large seed, when compared to a smaller seed, are less of a factor than the balance of nutrients within the seed and its physiological maturity. Our concern is whether the seed matured physiologically and received its full complement of nutrients.

The best, balanced, fertility package to which plants will be exposed comes from a physiologically mature seed which contains its full genetically programmed nutrient package. The seed's nutrient package is the "colustrum milk" for germination and early growth.

If given a choice we should select the first seeds to fill and mature on a plant. Our well-planned fertility program must not be diluted by either the construction of a poor seedbed or the selection of seed poor in quality. A good seedbed and top quality seed will set the stage for increased yields simply because two very important "yield limiting factors" have been eliminated. Everything that is done during and after planting must also complement these two most important areas.

ROW SUPPORT

"**Row support**" is a term coined by this author to designate a system designed to put into the row, directly onto the seed, a colostrum-milk-like package of materials for the purpose of increasing seedling vigor by overcoming and supplementing early seedling environmental deficiencies. **Row support** is a practice and the term does not relate to a specific product. The materials and nutrients used in the row may vary depending on the crop grown and local conditions.

Row support must be liquid because liquids offer a diversity of products, the ability for thorough blending of small amounts, and for uniformity and ease of application within the seed zone. The row support system can be liquid regardless of whether the starter system is liquid or dry. We should not think in terms of diverting part of the starter fertilizer into the row, but instead of a complete separate system.

This is because starter fertilizer and row support systems have two completely different objectives. The small quantities of materials

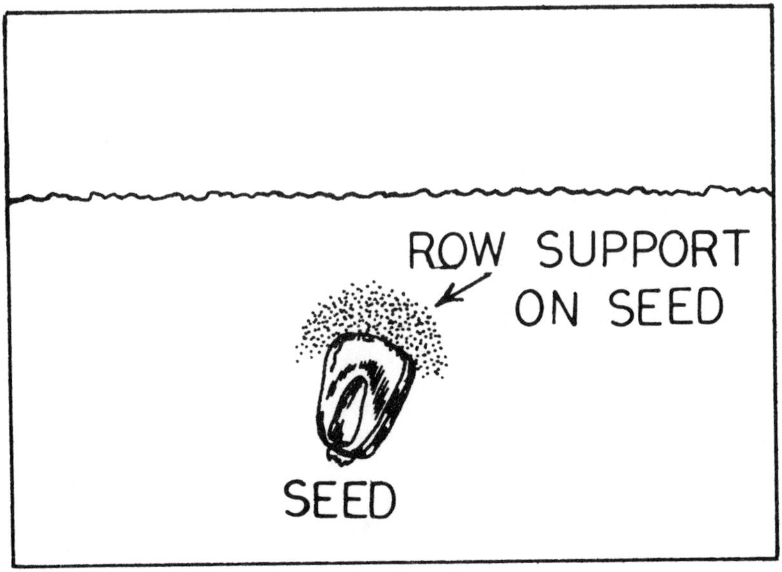

Fig. 125

placed on the seed would not be effective if placed off to the side in a normally positioned starter band.

Row Support is designed to improve the start and must be placed directly into the row with the seed as shown in Figure 125. At stated in its definition, the purpose of **row support** is to increase seedling vigor by overcoming and supplementing early seedling environmental deficiencies.

What are these environmental deficiencies? Weather conditions can change drastically, particularly during the early growth cycle. Soil conditions can change from warm and moist to cold and wet within hours during the germination and early growth periods. Cold soils retard microbiological activity which can influence soil temperature and nutrient release. Cold temperatures slow the speed of chemical reactions within the soil and living roots. Organic and inorganic nutrient release, availability and uptake are greatly reduced. Root growth and root functions slow down as temperatures drop. Such adverse conditions are not conducive to balanced plant nutrition and a healthy start. These are not the conditions under which we want to force vegetative growth with an excess of nitrogen.

Forcing vegetative growth under these adverse soil conditions will most certainly induce plant deficiencies. **Row support** offers the opportunity to apply a package of readily available nutrients and soil supportive materials to at least partially overcome the adverse environmentally induced deficiencies. Under poor soil and weather conditions the soil cannot serve up the nutrients needed, so we must. That is the real purpose of **row support.**

In Figure 126 we give recommendations for **row support** for four major crops. We recommend the row support system be calibrated to apply a total of six gallons per acre. The six gallons is based on

ROW SUPPORT RECOMMENDATIONS								
RECOMMENDATIONS BASED ON PER ACRE RATES FOR 28" TO 30" ROW SPACING								
CROP	10-34-0 POLY PHOS.	WATER	8-10% ZINC CHELATE	6% MANGANESE CHELATE	LIQUID INOCULANT	SURFACTANT	SUGAR	IRON CHELATE
CORN	6 GAL.	0	1 PINT	1 QT.	0	2-4 OZS.	0	1 PINT *
MILO	4 GAL.	2 GAL.	1 PINT	1 QT.	0	2-4 OZS.	0	1 PINT *
SOYBEANS	1 GAL.	5 GAL.	1 PINT	1 QT.	1 GAL. PER 30 ACRES	2-4 OZS.	½ LB.	1 PINT *
EDIBLE BEANS	0	6 GAL.	1 PINT	1 QT.	1 GAL. PER 30 ACRES	2-4 OZS.	½ LB.	1 PINT *
* IRON CHELATE MAY BE NECESSARY ON COLD, POORLY DRAINED OR HIGHLY ALKALINE SOILS.								

Fig. 126

our maximum recommendation of 10-34-0 for corn. Tolerance levels of nitrogen on the seed vary greatly depending upon the crop being grown. Wherever less than six gallons of 10-34-0 are recommended we recommend substituting water to make up the difference in volume. This permits the use of a uniform pump setting for all crops. Six gallons provides a reasonable volume while minimizing tank capacity.

The suggested ingredients in Figure 126 contain both **nutrients and soil supportive materials.** The nutrients can be taken up directly by the roots. The soil supportive materials are designed to improve the soil environment in relationship to water management, biological activity, nutrient release, etc. Each ingredient has a specific purpose.

First, you will observe that we are recommending 10-34-0 polyphosphate. There are several reasons for this recommendation. A good grade of polyphosphate will contain about seventy percent of its total phosphorous in the poly form and the balance will be in the ortho form. These forms of phosphorous are explained in the Phosphorous section of this book.

The polyphosphate has definite advantages over the orthophosphate. It resists tieup with other nutrients and releases or changes into the high-available ortho form when needed. Orthophosphate is the form most used by plants and is also the form most readily tied up with other elements within the soil. Orthophosphate may become locked up within hours after application. The rate of conversion from polyphosphate to orthophosphate depends on soil conditions. Soil conditions favoring germination and early growth are the same that favor soil microbiological activity and the conversion of poly to the ortho form. It is a slow release process that coincides with conditions favoring germination and early root development. The thirty percent ortho form may remain available or soil conditions may cause it to become locked up. By locking up it will tend to neutralize the soil lockup agents. As the polyphosphate converts to orthophosphate it will be more available because of this early neutralizing action. The use of polyphosphate can nearly guarantee phosphate availability during the critical early seedling stage.

A limit of six gallons per acre of 10-34-0 has been placed for corn. We believe this to be near the maximum amount that can be effectively utilized by the crop. Most fertilizers have a salt index indicating salt action that can cause fertilizer burn. Nitrogen and potash are two ingredients that can cause fertilizer burn and our strict rule is to limit the amount of nitrogen, or nitrogen with potassium, to a maximum of eight pounds per acre on the seed for corn. Corn has a much higher tolerance than many other crops.

The six gallon recommendation supplies only six pounds of actual N per acre which is well under the eight pound limit for corn. Many farmers use ten gallons or more of 10-34-0 per acre on corn. This is very risky and there seems to be a definite limit on the rate to use for maximum effect. By using **row support** we **are not** trying to fertilize the entire crop, but simply to give it a good start.

Potassium has purposely been excluded from the row support because we believe phosphorous is more important during the early seedling stage. Our purpose with **row support** is to grow roots first before an attempt is made to accelerate the top growth. By the time a corn plant reaches a height of four inches the roots should outweigh the parts above ground. We want to produce a massive root system under the small plants.

Plant-usable phosphorous is essential for root growth and branching. When potassium is included in a liquid fertilizer, such as 6-18-6 or 7-22-5, a considerable amount of phosphorous is sacrificed in comparison to a 10-34-0 formula. We do not believe adding potassium is worth the sacrifice of phosphorous especially while keeping in mind that we do not recommend exceeding the eight pounds per acre of nitrogen or a combination of nitrogen with potassium. The potassium blends are too limited in phosphorous for our row support goals.

Our recommendations for milo are for four gallons of the 10-34-0 simply because we know that four gallons is a safe level and works well. Our experience in using the higher rates on milo is limited, so we must play it safe.

Soybeans do well with one gallon of 10-34-0, but might have problems at the two to three-gallon level. Some farmers have used three gallons and gotten along very well and were pleased with the results. However, when it takes soybeans ten to fifteen days to emerge because of dry soil conditions the two or three-gallon rates have been known to thin the stand.

The other recommended items in the row support are more important for the soybean plant than the small amounts of nutrients in the one gallon of 10-34-0. The one gallon of 10-34-0 is included mainly to provide a little nitrogen and phosphorous for the nodular nitrogen-fixing rhizobia bacteria. This along with the carbon energy supplied by the sugar appears to stimulate early nodulation on the bean roots.

We consider the zinc chelate to be very important even though the general soil levels for zinc are in the high range. A small amount of zinc in a highly-available chelate form is very important at the start because of the many important roles zinc plays in the developing plant. It is believed that zinc plays an important role in phosphorous uptake. Because zinc is essential to at least thirty and possibly as many as fifty enzyme actions within the plant cells it is vital to nearly every life process, which is why we consider the addition of a highly-available source in the **row support** so important.

Manganese is particularly important for photosythesis and its availability is very poor under cold, wet soil conditions. As the soils warm and roots expand manganese becomes adequate on most soils. Our concern is to assure manganese availability under adverse soil conditions to prevent the first leaves from developing visual or hidden chlorosis which reduces their ability to produce sugar through photosynthesis. Root growth will slow or even stop without an energy supply from above. A slight deficiency in manganese could greatly reduce root growth and limit the benefits of row support.

Our general guidelines on the use of manganese reflect both soil conditions at planting time and also geographic location. Regardless of the geographic location we always include manganese in the row support if the soil manganese levels are low or marginal. If soil levels of manganese are adequate it becomes less important as we move into southern climates with warmer springs and longer growing conditions. The farther north we go with the shorter growing seasons the more important it becomes to automatically include manganese in the **row support.** This also applies to poorly-drained soils. In areas with longer growing seasons we add it only when conditions are poor or unfavorable for good germination and early growth.

While all of the trace elements are important zinc and manganese seem to be most necessary for early growth. Boron is too toxic to attempt to regulate when in contact with the seed. Copper is risky and may cause a deficiency of manganese if it is in excess. Iron could be beneficial at one pint of chelate per acre, particularly on cold, poorly-drained, or alkaline soils. Under these conditions tests should be run using one pint of iron chelate per acre to observe any resulting benefits. An iron deficiency will also cause leaf chlorosis and reduced sugar production. As with manganese, iron deficiency is most likely to occur early in the season when adverse soil conditions may hinder its release.

For soybeans the liquid inoculant provides a convenient method of dispersing a high concentration of rhizobia bacteria within the seed zone. At the rate recommended in Figure 126, you will be applying more organisms per acre than with the standard method of inoculating the seed. The nutrients being applied at the recommended levels will not harm the rhizobia bacteria in the inoculant.

One-half pound of sugar per acre has the purpose of providing a ready source of carbon energy for the rhizobia and other organisms within the seed zone. Small amounts of sugar seem to stimulate nodules to form on the roots much earlier than when it is not used. These organisms require carbon energy which the sugar provides with or without inoculants. Our position is that if early nodulation is good, the use of sugar is a good practice. It must be pre-dissolved before its addition to the row support tank mix. Sugar is recommended because of its readily available supply. Other products, such as feed grade molasses at one pint per acre, may be superior to white sugar because of the other minor ingredients it contains in addition to the carbon.

We encourage farmers to set up tests with and without using sugar and then comparing the root nodulation from the time the plants reach three to four inches high. Check them at weekly intervals and, if feasible, check the yields. Tests should be done using sugar with and without inoculation on land that has had several crops of beans in the past. We must first build a good soil environment before inserting any kind of beneficial soil organisms, including rhizobia carrying inoculants. Sugar is simply a means of improving the microbiological food supply within the soil environment. The inoculants and sugar are soil supportive materials.

The surfactant recommended is also soil supportive and can help manage water within the seed zone. Surfactants may speed germination by their influence on water adsorption by the seed. They can also serve as a mild nutrient-releasing agent by increasing nutrients in the soil solution and making them more accessible and available for uptake by the young roots.

The recommended use of surfactants is at very small rates confined to the seed zone. This is the "action zone" or the place where we want things to happen to improve the welfare of the plant.

We must be concerned with the heavy broadcast recommendations and claims being made for some surfactants. Some even claim fertilizers can nearly be eliminated by broadcasting high rates of their surfactants. Surfactants do not contain any appreciable amount of plant nutrients. They may be used for one or two years without adding anything except nitrogen with good results, but this could be done on many fertile soils even without a surfactant. Such thinking has left a trail of wreckage across the countryside. Our major concern with such practices is the lack of research and knowledge as to exactly how these products function within the soil. They may free up sufficient quantities of nutrients so that fertilizers are not necessary for awhile. However, what affect will this "freeing up" process have on the leachability of soil nutrients? Crop failure may suddenly result from the mass exodus of some nutrient, such as boron, which is already borderline on many soils. It would require the rinsing out of only one essential nutrient to cause a crop failure.

We think, with reasonable certainty, that one to four ounces of surfactant placed within the seed zone can enhance the early growth process with little or no threat to the soil system. From our own observation we believe that surfactants within the seed zone, as recommended, are most beneficial on heavier-textured soils. We have seen little benefit when it was used on sand or sandy loam soils. On heavy-textured soils and other less than ideal soil conditions we have observed a soybean height of six inches where four ounces of surfactant per acre was used as **row support,** in comparison with a height of four inches where the surfactant was not used. You may not see such visual results under more ideal soil conditions.

The **row support** system offers much potential for enhancing the early growth period of a plant. Microbiological activity within the root zone or rhizosphere seems to be necessary to the welfare of the plant. There are symbiotic reactions between specific organisms working within the rhizosphere of each species of plant that are necessary for the welfare of that particular plant. For example, there are certain organisms that thrive on the roots of trees that are necessary for the tree's survival. This symbiotic soil organism-plant relationship may apply to all plants.

Row support offers a method of applying soil supportive materials that can enhance the microbiological environment within the root zone. In the near future we may even be able to insert certain organisms within the seed zone that are helpful to the specific crop being grown. We may be able to add chemicals such as calcium peroxide (CaO_2) to release both calcium and oxygen. The soil environment and soil organisms must be compatable and **row support** offers a simple means to favorably alter this environment.

Many organic products are presently on the market that may be best used when applied in small quantities directly into the row. Any experimentation in this respect should be done strictly on a limited basis.

The seed heads of milo form within the plant during the first two or three weeks of growth. Corn ovulates and determines the number of rows of grain on the cob about five weeks after emergence. The genetic potential is set very early in the life cycle of many crops. The purpose of **row support** is simply to maximize this genetic potential. What a difference two extra rows of corn on each cob can make in profits!

A good seedbed, strong, viable seed and **row support** are three areas of the "start system" that can greatly enhance profits. Figure 127 shows the tremendous difference a properly managed **row support** system can make in corn. The shorter corn on the left did not receive row support nutrients at planting time. Figure 128 shows the difference **row support** can make in a season with adverse weather conditions.

Fig. 127

The corn treated with **row support** will be more vigorous in both topgrowth and root growth. A large, vigorous root system and hardy tops as shown in Figure 128A can be grown providing the right nutrients are supplied within the seed zone. Again, it must be emphasized, we do not want to grow tops at the expense of roots. If top growth is forced with an excess of nitrogen alone it will almost certainly weaken the plant and reduce its genetic expression.

Fig. 128

Fig. 128A

Row support will minimize the loss of growth time early in the season and keep uniformity in plant growth. It will strengthen the weaker seeds and help prevent them from falling behind the others in growth. A shorter plant simply cannot compete with the taller ones and will not produce as well.

Notice the differences in the growth of corn, as shown in Figure 129. This corn was all planted on the same day. The taller, tasseled corn has gained several days in growing time compared to the slower developing corn that is not tasseling. This unevenness can be prevented or greatly minimized with a properly managed **row support** system.

Early stress, as shown in Figure 130, must be avoided. Purple coloration can result from phosphorous deficiency caused by

Fig. 129

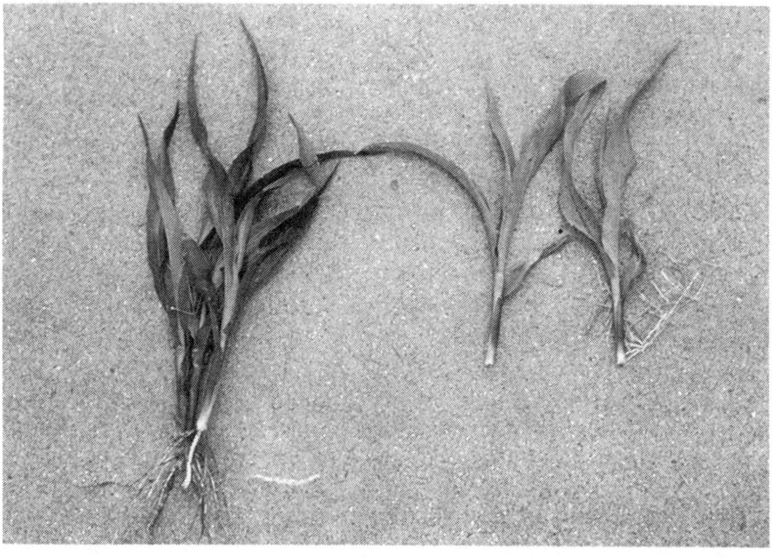

Fig. 130

prolonged cool, moist soil conditions. It is believed that other stress factors may also cause the loss of green chlorophyll and the exposure of the purple pigment. Regardless of the cause, the loss of chlorophyll is always detrimental to the welfare of a plant.

Row support will assure a reasonable balance of nutrients during early growth that will place emphasis on developing roots before tops. A large, expansive hunt-and-gather system under a plant is the best insurance for delivering a balanced diet to the developing plants, providing of course, that the soil itself has been balanced. A really professional farmer would not prepare an ideal seedbed, select top quality seed, use **row support,** and then watch it all fail later in the season because of a major deficiency of some other nutrient, such as boron. The professional farmer would have identified that problem and taken the necessary steps to correct it before he planted the crop.

The equipment required to apply **row support** is relatively simple and easily set up. Figure 130A is a schematic drawing illustrating

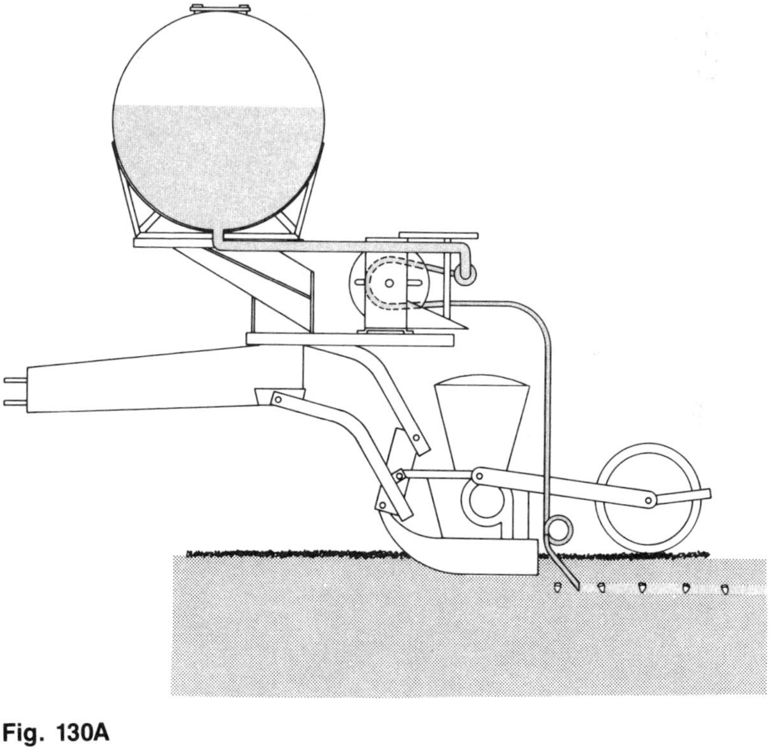

Fig. 130A

a simple setup using a ground-driven metering system, most often called a "squeeze" or "roller" pump. This system is very accurate regardless of ground speed. The tank capacity need not be too large since only six gallons per acre are being applied. A one hundred twenty gallon tank would supply enough material for twenty acres. This will usually more than match the supply of seed or starter fertilizer that can be carried.

Figure 131 shows an electric pump setup to supply **row support** on soybeans. Figure 132 illustrates a combination of electric pump and squeeze pumps. The small electric pump in the background supplies the **row support** and the squeeze or roller pumps in the foreground supply the starter fertilizer and nitrogen on this eight-row planter. Figure 133 shows a squeeze pump setup to supply **row support.**

The electric pump kit shown in Figure 134 is a very economical and simplified method of providing **row support.** These pumps are very durable, low in cost and simple to install. They can be easily removed and transferred to the cultivator, as shown in Figure 135 and 135A. Their output is simply controlled by an on-off switch and a motor speed controller conveniently mounted in the cab. These kits are ordered according to the number of rows on a planter. Their row capacity can be increased by adding two-row increments to the outlet manifold. The pump has the capacity to supply a maxium of thirty-four outlets. In this respect it would adapt to grain drills. These electric pumps can handle a maximum of twelve rows using 10-34-0 at six gallons per acre.

Electric or roller pumps are the ideal systems for **row support.** Do not attempt to set up a gravity-feed system because they will not accurately meter materials as the supply in the tank drops. Also, do not try to provide **row support** nutrients by metering part of the starter fertilizer into the row. **Row support** must be a separate system because its objectives are entirely different from the starter fertilizer principles.

By preparing a good seedbed, selecting vigorous seed and using **row suppport** you will be well on the way to success. Everything possible will have been done to establish the start phase of the total production system. Prior to this, of course, the basics of soil aeration, soil water and residue decay will have been carefully managed. Limestone will have been spread if needed and broadcast phases of fertility will be accomplished.

The final phase of the fertility program's implementation will be inserting the starter and nitrogen systems where they are needed. Starter fertilizer may not always be needed depending on the soil situation and the crops to be grown. Nitrogen systems will also vary depending upon the crops to be grown. Nitrogen management for growing corn will certainly differ from that for soybeans. We must

Fig. 131

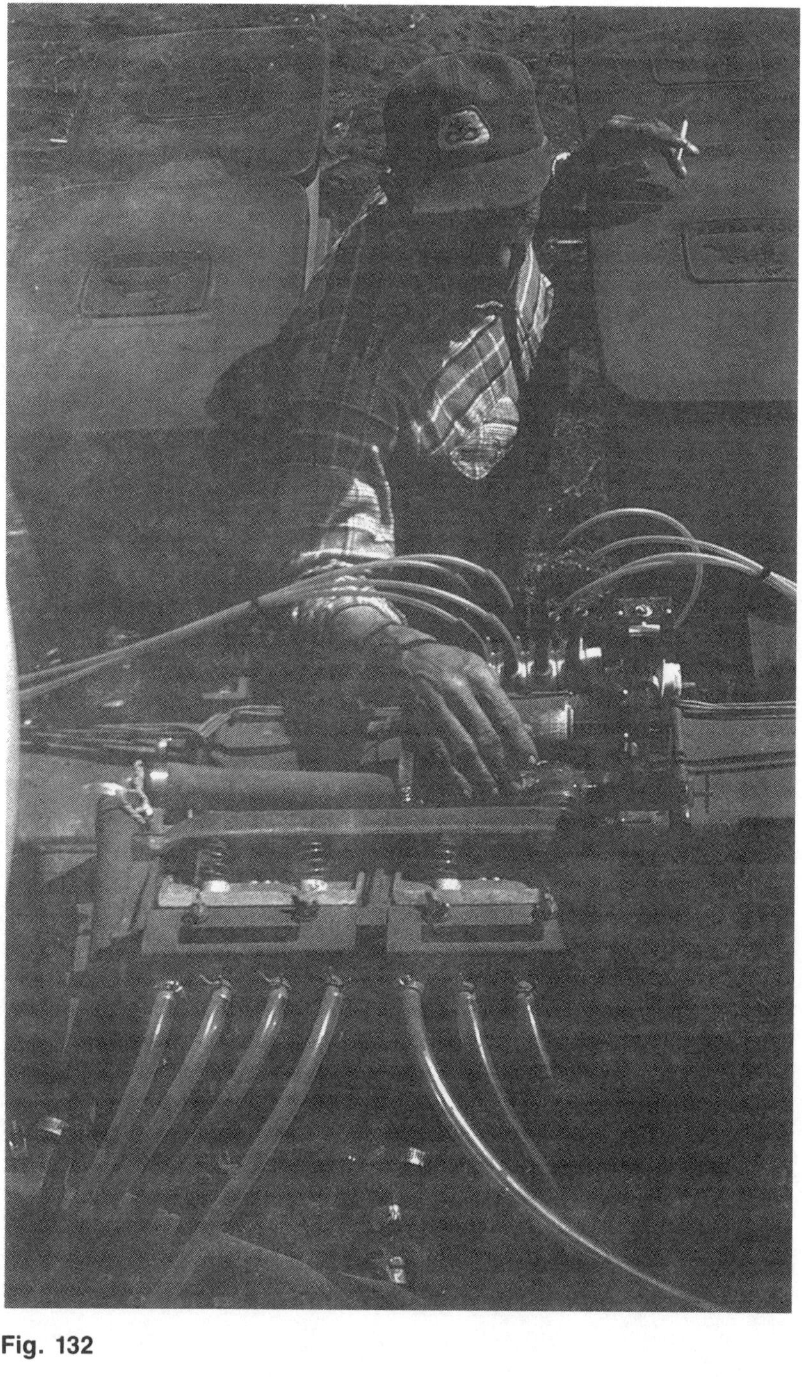

Fig. 132

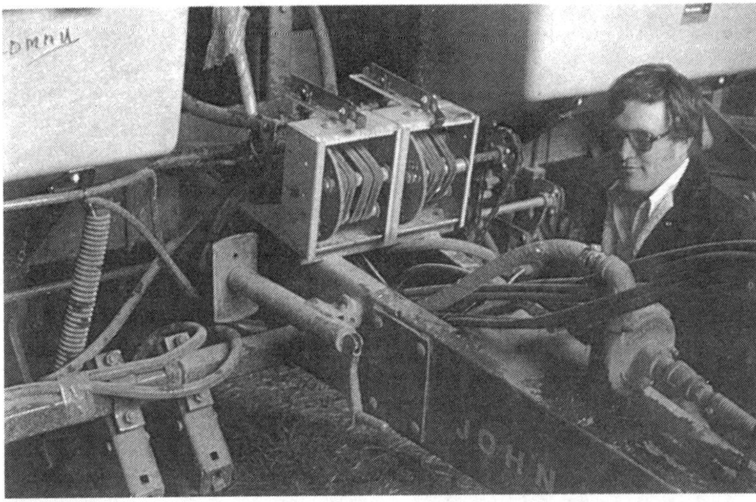

Fig. 133

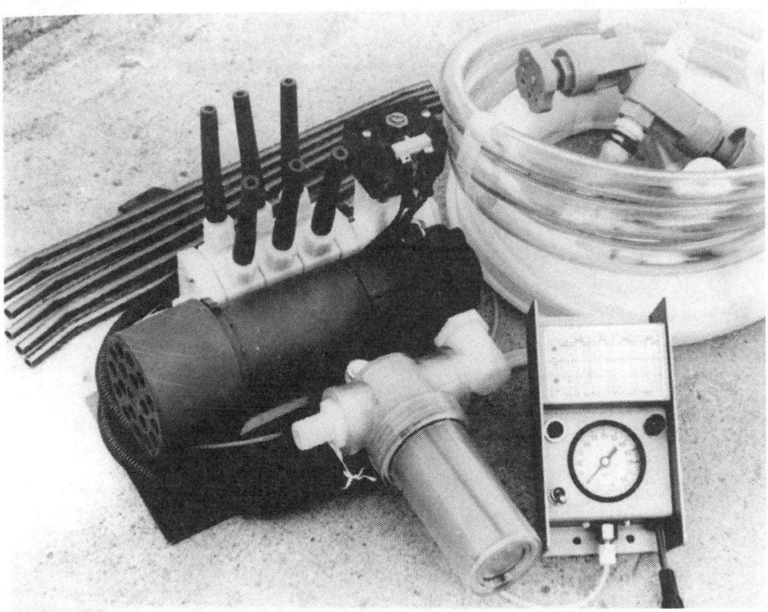

Fig. 134

study nitrogen and starter fertilizer management in relationship to specific crops.

The high nitrogen-demanding crops of corn and milo will be first in our discussion. The principles applying to them also apply to other

Fig. 135

Fig. 135A

crops such as sunflowers and cotton. We are more interested with the basic principles rather than specific rates per acre for the elements such as nitrogen or phosphorous at this time.

STARTER FERTILIZER AND NITROGEN SYSTEMS

With the **row support** system in use we will have assured a good start. The **row support** provides options that we would not otherwise have. With its use nitrogen can be concentrated closer to the plant without fear of overstimulating top growth at the expense of early root growth and nutrient balance.

Placement of nitrogen in a concentrated band near the plant offers the best opportunity to drastically cut costs and actually increase yields in the process. To fully appreciate the placement of nitrogen we must understand some basic nitrogen chemistry and principles relating to placement. Therefore, we will study nitrogen alone, as well as the reasons for its application in combination with other nutrients.

First, we will view the root system of a corn plant, as shown in Figure 136. Imagine the root mass as being situated in a bowl, which we will call the "**zone of updraft.**" Visualize water moving into this "zone of updraft" from all directions to replace the water being taken up by the root mass. The movement of the water toward the root mass is significant in the flow or direction of nutrient movement. Nutrients are actually carried into the "zone of updraft" with the flow of water.

A root takes its nutrients directly from the soil particle or from the capillary moisture film surrounding the soil particle.

As mentioned earlier, the cations on the negative soil particle tend to maintain an equilibrium with the cations in the moisture film surrounding it. This is illustrated in Figure 136A, showing the relationship of storehoused and exchangeable potassium to the potassium in the moisture film surrounding a clay particle. This principle also applies to other nutrients.

As water evaporates from the soil surface the capillary film continues to move upward, carrying nutrients and depositing them on or near the soil surface. This principle also applies to the "zone of updraft." As the root mass takes in water, the replacement water moving in from all directions sets up a mass flow of nutrients into the root zone.

An interesting fact to note is that nutrient levels tend to equalize from one moisture film to another as well as between soil particle surfaces. If a single clay particle was stripped clean of its cations it would not remain barren. Because of its tendency to equalize nutrients would move from the soil particles with the higher concentrations to the barren particle until an equilibrium was established.

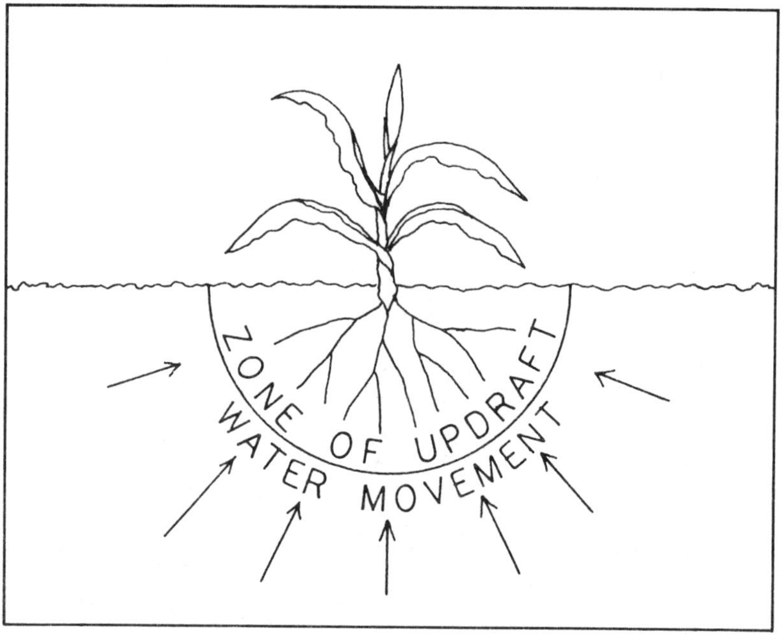

Fig. 136

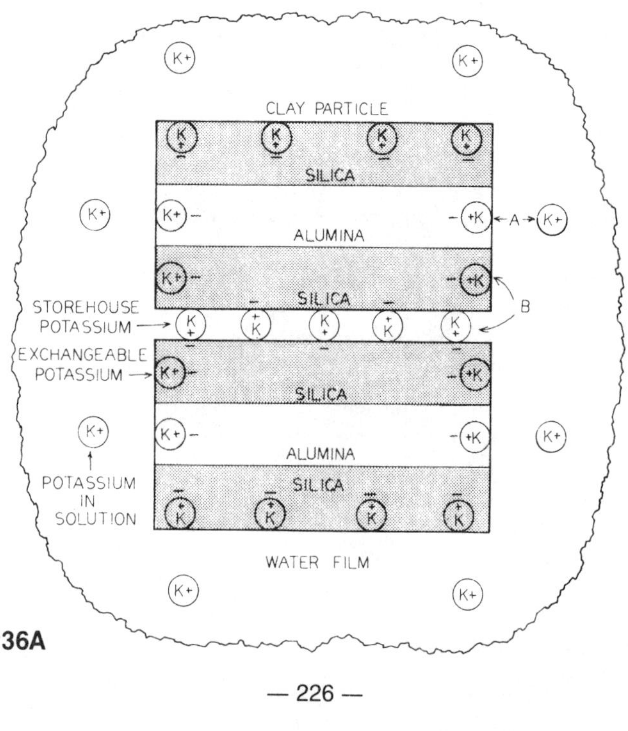

Fig. 136A

This nutrient movement would take place through the moisture film that interconnects all of the soil particles. The "zone of updraft" principle is significant as it relates to the mass flow of nutrients into the root mass. This principle is particularly meaningful in the management of nitrogen and makes clear the reasons for placing it within the zone of updraft.

Nitrogen's major role is in the formation of amino acids and proteins. Proteins are essential to life. When life ceases a host of organisms immediately go to work to break down the proteins and release the nitrogen into plant usable forms.

This recycling of nitrogen is accomplished through three separate processes known as aminization, ammonification and nitrification. These are shown in Figures 137, 138 and 139.

Fig. 137

AMINIZATION

BREAKDOWN OF PROTEIN TO AMINO ACIDS

$$\text{PROTEIN} + \xrightarrow[\text{AND FUNGI}]{\text{SOIL BACTERIA}} \text{AMINO ACIDS}$$

Fig. 138

AMMONIFICATION

BREAKDOWN OF AMINO ACIDS TO AMMONIA

$$\text{AMINO ACIDS} + \xrightarrow[\text{AND FUNGI}]{\text{SOIL BACTERIA}} \text{AMMONIA } (NH_3)$$

Fig. 139

NITRIFICATION

CONVERSION OF AMMONIA TO NITRATE

$$NH_4 + \xrightarrow[\text{BACTERIA}]{\text{NITROSOMONA}} NO_2 + \xrightarrow[\text{BACTERIA}]{\text{NITROBACTER}} NO_3 + 4H$$

Aminization is the process of breaking down proteins into the individual amino acids that comprise it. The **ammonification** process degrades or breaks down the amino acids and releases ammonia. Aminization and ammonification are done by several different soil bacteria and fungi. The final stage of **nitrification** is the very important step of converting ammonium to nitrate.

Figure 140 shows why nitrification is so important to nitrogen placement. As shown, ammonium is positive-charged (+) before being

NITRIFICATION

$$\underset{\text{AMM.}}{NH_4^+} + \xrightarrow[\text{BACTERIA}]{\text{NITROSOMONA}} \underset{\text{NITRITE}}{NO_2\uparrow} + \xrightarrow[\text{BACTERIA}]{\text{NITROBACTER}} \underset{\text{NITRATE}}{NO_3^{=}\downarrow} + 4H^+$$

Fig. 140

changed to nitrite. We show nitrite with an arrow pointing upward, indicating that N, NO and NO_2 are gasses, and as such, can escape from the soil.

The conversion from ammonium to nitrite is brought about by **nitrosomona bacteria.** The nitrite is then converted by **nitrobacter bacteria** to nitrite, which is negative-charged (−). In other words, nitrogen starts out as a cation (+), but ends as an anion (−).

As illustrated in Figure 141, a soil particle is negative-charged and the positive-charged ammonium will attach to it, while the negative-charged nitrate will be repelled by the negative-charged soil particle.

The nitrate molecule has the same negative charge as the soil particles. Since like charges repel the nitrate has no home in the soil and must move with the water. Nitrates can move downward and may be lost through tile drainage during periods of excessive rainfall. Nitrates can also move upward and accumulate at or near the surface during extended dry periods.

Since all nitrogen eventually ends as negative-charged nitrate (NO_3=) that readily moves with the soil water it is immediately apparent that proper placement within the "**zone of updraft**" should significantly reduce nitrogen losses.

The key to reducing nitrogen costs in production of a crop such as corn is to reduce nitrogen losses and to increase nitrate recovery.

Nitrogen may be the most abused and misused production input in American agriculture. To appreciate this statement it will be necessary to quickly review the present standards for nitrogen use.

The present nitrogen recommendations for growing a given yield of corn are based on broadcasting and incorporating nitrogen before planting the crop. As a result a common practice has come about of mixing herbicides with the nitrogen and broadcasting both together in a "weed and feed" operation. Broadcasting nitrogen in a weed and feed approach may save a trip over the field, but it is the most wasteful and humus-destroying system imaginable. If farmers had

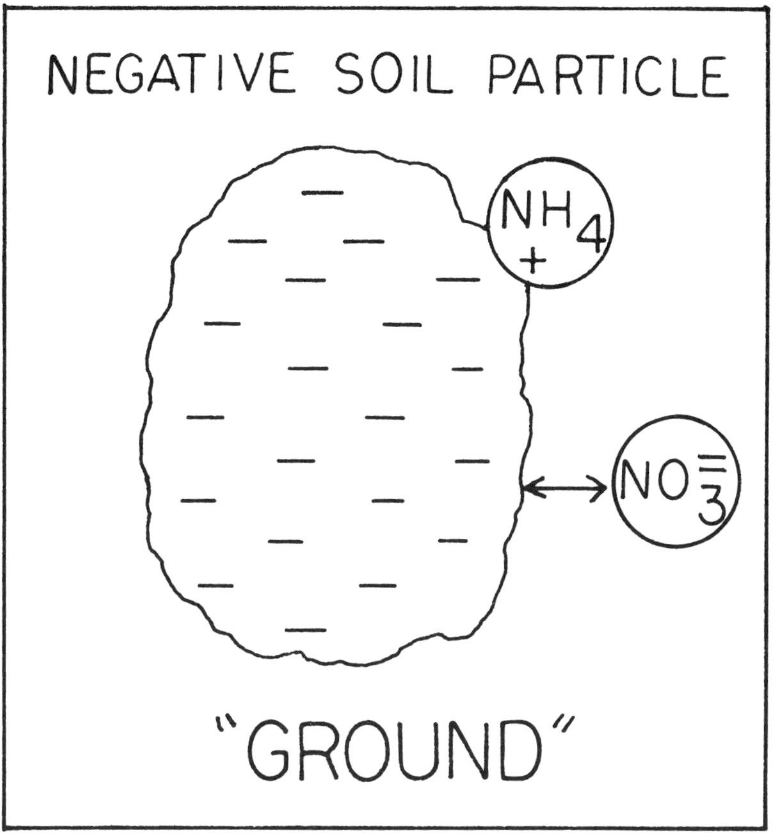

Fig. 141

the money in their pockets and the tilth back into their soils that this system has cost them over a twenty-year period it could possibly buy a sizeable portion of their farms. In addition broadcast nitrogen may actually stimulate the growth of weeds.

We are convinced that a crop such as corn does not recover much more than fifty percent of nitrogen that is broadcast prior to planting on light or heavy textured soils. What accounts for this huge loss of broadcast nitrogen?

To understand these losses we must review the nitrification process illustrated in Figure 140 one more time. As ammonium moves toward the nitrite form it becomes a gas. The losses from the soil as a nitrogen gas are much greater than previously thought. The nitrate form can leach from light-textured soils and combine with other elements such as potassium and calcium or accumulate near the surface during dry periods. Wet conditions tend to move nitrates downward and dry conditions move it to the surface.

Nitrogen becomes locked up in the protein bodies of soil organisms and tied up with soil organic matter. It may also become trapped in the wheel tracks. Nitrates can be reduced back into the gaseous form by anerobic soil organisms in soils that become low in oxygen. The losses from gasification will be more on heavy soils than on light-textured soils. Leaching losses will be higher on light soils.

The combined losses are staggering and could exceed fifty percent. Some may argue this, but field testing will soon convince the most skeptical that there is something drastically wrong with the present recommendations which are given for nitrogen. Under a broadcast, pre-plant, nitrogen application system these recommendations could be correct. However, it is time to review this practice with the realization that it is based upon the loss of at least forty percent of the nitrogen. This loss is unacceptable.

How does the practice of broadcasting excessive nitrogen destroy the soil humus and tilth? Soil decay organisms require nitrogen to build their protein bodies. When excessive nitrogen is present in the soil these organisms will multiply if they have access to a source of carbon for energy. By bathing the soil with an excess of nitrogen the soil organisms will have easy access to it. In addition to using carbonaceous residue the soil organisms can utilize humus as a source of carbon. Since humus is uniformly distributed throughout the soil it is more accessible to the soil organisms than the randomly distributed crop residue. As a result of the uniform dispersion of nitrogen and humus the soil organisms will utilize or break down the humus for their carbon needs. This is how the application of excessive nitrogen that goes beyond the needs of the crop will deplete humus.

The ideal method to decay carbonaceous residue such as cornstalks is to spray a liquid nitrogen directly on the plant prior to incorporation. This minimizes the bathing of the entire soil with too much nitrogen.

We must recognize the potential for heavy nitrogen losses when we apply it improperly. For the moment we will concern ourselves with the basic principles of nitrogen management rather than the rates needed to produce a given yield. Rates of application will be reviewed at the end of this unit.

These principles and methods of nitrogen application in combination with other nutrients are illustrated in Figure 142. This illustration shows the **row support** on the seed, the starter fertilizer band placed off to one side and a nutrient band placed alongside the plant

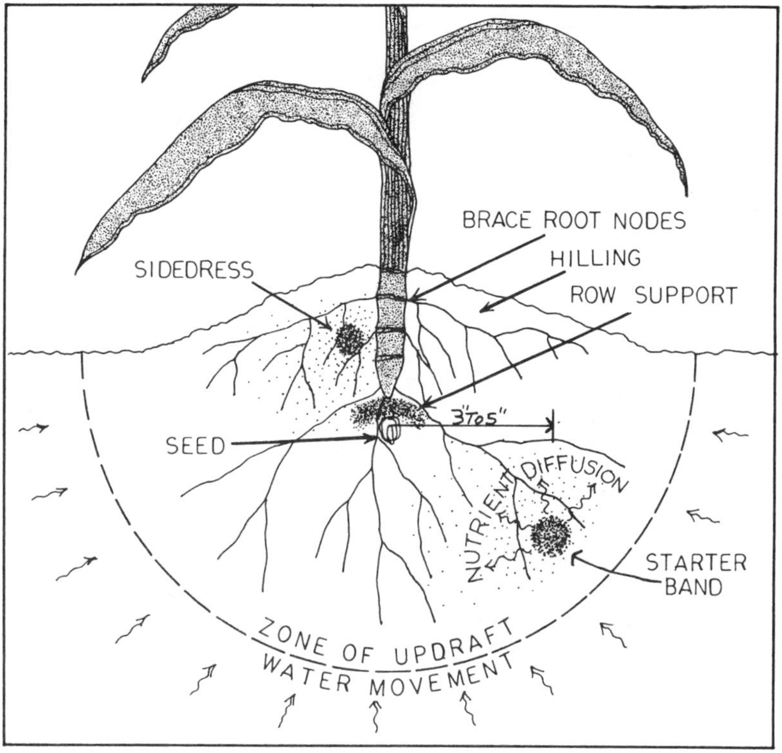

Fig. 142

during cultivation. Figure 142A shows the excellent early root proliferation from the row support and starter bands.

The starter fertilizer band is placed out three to five inches to the side of the row. There are important reasons for doing this. First, by moving it out three to five inches to the side of the row we may use all the nitrogen needed without fear of harming germination. Second, the nitrogen can be set out to minimize early "luxury feeding" of nitrogen which could force top growth at the expense of early root growth. You will immediately realize that this is made possible because of the use of **row support** to manage the early growth. The plant does not need to depend on the starter band for its early needs.

You will see in Figure 142 that the nutrients are diffusing out of the starter band and moving into the root mass. Since water is moving into the **"zone of updraft"** from all directions, the nitrates and other nutrients tend to be trapped within the root mass. This is ideal because of the improved nutrient accessibility, availability and recovery. As mentioned earlier, soil aeration is better within the root

Fig. 142A

mass than in any other part of the soil beyond the root mass influence. We expect better recovery of nutrients within the "**zone of updraft**" because of root access to a more plentiful supply of oxygen.

Recommendations have been made for **row support** nutrients and it has been established that **row support** must be in a liquid form. The next logical questions may be whether the starter fertilizer should be liquid or dry, and what nutrients should be included in the band.

"Starter fertilizer" may be poor terminology when referring to this side-placed band of nutrients. Since it is understood that this band is not being used for the sole purpose of starting the plant's growth we will not confuse the issue by introducing new terminology. This band, as we shall see, may supply the entire nitrogen needs of the crop or even provide maintenance levels for other nutrients such as phosphorous.

The starter fertilizer band may be either liquid or dry depending upon your personal preference, forms available in your own community, and present equipment on hand. Our own personal preference is an all liquid system rather than a dry. Reasons for this are because liquids are easier to handle, blending of ingredients is more precise and uniform, blending can be done by the farmer himself and the **row support** will be in the same form as the sidedress.

There are not appreciable differences in nutrient availability between the liquid and dry applications. Dry forms may have a slight edge because the phosphates may carry a little calcium, or we may argue that dry, sulphated trace elements are less expensive in dry starters, as opposed to the chelates used in liquids. The fact remains that the selection between a liquid or dry system has less to do with N, P or K availability and more to do with unit costs of nutrients, suppliers in an area, ease of handling and personal preference.

The depth at which the starter band is placed is an important consideration. Soil moisture is necessary for root function, nutrient solubility and recovery. The diffusion outward from the starter fertilizer band is much slower for phosphorous and the cations, such as potassium and trace elements, then it is for the negative-charged nitrate. Nitrate is mobile and moves readily through the moisture film. The less mobile plant nutrients are more dependent upon an adequate moisture supply to remain available for uptake by the root system.

Roots must have moisture to function. When the soil dries to the wilting point, the root hairs die and the nutrient uptake ceases or is greatly reduced. When moisture returns to the soil new root hairs must grow before sufficient nutrient absorption can resume.

We believe that the standard settings for starter fertilizer are simply not deep enough to assure continuous and maximum recovery of nutrients from the starter band. Keep in mind that the use of **row support** permits the placing of the starter band both farther out from the row and deeper. **Row support** minimizes an early dependence on the starter band.

The depth of the band is determined by soil type and rainfall patterns in the area. Greater depths are necessary in semi-dry areas than in irrigated soils or areas of abundant rainfall. Where there is normally enough rainfall the starter bands should be placed at a minimum of four inches deep. The applicator should have the flexibility to go as deep as six inches if necessary. The band should be placed deeper on sandy soils than on high moisture-holding, organic soils. In fact, a four to five inch depth throughout the midwest would be an improvement on the standards presently in use. Again, we remind ourselves that this is made possible through the use of **row support.**

The capability of adjusting the applicators out three to five inches or more from the row not only prevents early luxury feeding of nitrogen but disturbs less soil near the seed. More soil will be disturbed as depth is increased.

Figure 142 also shows a sidedress band placed on the surface near the base of the plant while cultivating. You will notice that this band has been covered in the hilling process while cultivating.

In our unit on cultivation we mentioned that a cultivator should be equipped to apply a nutrient and lay up a hill around the base of the plant. If a cultivator is not designed to lay up a hill the band could be applied behind the tine or shovel nearest to the row. It should not be applied in the middle of the row. Another detail to notice in Figure 142 is that the sidedress band is on the opposite side from the starter band. The significance of this will be brought out a little later.

Figure 142 clearly shows the entire system for managing nutrients within the "**zone of updraft.**" The row support and starter band are the two most important components of the system. The sidedress at cultivation is optional in most cases but definitely necessary in some others. We need options and have several in this pattern of placement.

Our next area of concern relates to nutrient use within these bands. The **row support** will be a standard blend of materials dependent upon the crop being grown. Soil test results are the most important factor in determining nutrient needs within the starter and sidedress bands. There are also other considerations that must be taken into account to maximize results from this system.

Let us begin with phosphorous management. As you will recall, phosphorous is very vulnerable to becoming locked up by many factors within the soil. Under certain conditions, such as high alkalinity, low pH, low calcium or extremely high magnesium, it may be absolutely necessary that all of your phosphorous needs be banded to minimize exposure to these "lock up" conditions. You must be able to recognize these conditions as you study your soil test data.

Maintenance levels of phosphorous can be applied within this placement system regardless of soil conditions. Applications of estimated maintenance levels within this zone of updraft are desirable for several reasons. The soil will have a "full tank" at the start and will end with a full tank after harvest. This approach seems far superior to ending the season with a partial fill of phosphorous. Applications of maintenance levels of phosphorous at the beginning instead of replacement at the end of harvest assures an adequate supply of phosphorous even under adverse weather conditions, limited soil aeration, etc. Banding a maintenance level of phosphorous is desirable, particularly the first year after an application of limestone.

Maintenance needs for a given crop, such as corn, would be divided between the row support sidedress and starter band. If no sidedressing was done it would be divided between the row support and starter. For instance, with corn we would already have twenty pounds of P_2O_5 in the row support from the six gallons of 10-34-0. The amount of nutrients removed by various crops are shown in Figure 143.

Our attention will now switch briefly to the sidedress band, placed during cultivation, and its relationship to phosphorous management.

Figure 142 illustrates a hill that was laid up around the plant during cultivation. A hill can serve several functions. In a way it is similar to planting on a small ridge except the cultivator ridge is laid up post-plant. The hill helps to control soil aeration and minimizes the affects of excessive water.

Hilling is very important in growing corn as Figure 142 illustrates, because it will cover corn nodes and encourage them to shoot out an extra set of roots underground. The brace roots would then come off a higher node. This upper root development and underground proliferation is significant because it is believed that these roots feed most of the phosphorous to the upper part of the corn plant late in the season. (See Figure 7 on Page 30). An adequate phosphorous supply to the upper part of the corn plant late in the season is important for the grain to physiologically mature with good test weight.

You will notice that the sidedress is placed on the opposite side from the starter. There is a specific reason for this with corn. On corn the brace root buds emerge very close together in a ring completely encircling the stalk, as shown on Page 132. Each root bud

POUNDS OF NUTRIENT USE
and
REMOVAL BY VARIOUS CROPS

Cereal/Bu.	N Nitrogen	P205 Phosphorous	K20 Potassium	Mg Magnesium	Ca Calcium	S Sulfur
Corn	1.0	.35	.25	.06	.02	.07
Stover	.5	.25	1.05	.14	.19	.09
Total	1.5	.60	1.30	.20	.21	.16
Soybeans	4.2	.9	1.5	.23	.2	.20
Straw	1.3	.3	.9	.22	1.5	.25
Total	5.5	1.2	2.4	.45	1.7	.45
Wheat	1.6	.6	.4	.14	.05	.10
Straw	.5	.12	1.2	.10	.21	.13
Total	2.1	.72	1.6	.24	.26	.23
Milo	.85	.40	.25	.08	.07	.09
Stover	.85	.30	1.40	.15	.31	.10
Total	1.70	.70	1.65	.23	.38	.19
Oats	.7	.25	.2	.04	.03	.07
Straw	.35	.18	1.1	.11	.12	.11
Total	1.05	.43	1.3	.15	.15	.18
Barley	1.0	.4	.35	.07	.04	.08
Straw	.4	.15	1.10	.08	.20	.10
Total	1.4	.55	1.45	.15	.24	.18
Forages/Ton						
Alfalfa	60	14	45	5	28	6
Brome Gr.	40	12	40	4	8	4
Red Clover	55	12	45	6	24	5
St. Clover	44	11	45	5	29	8
Timothy	35	13	50	4	8	4

Fig. 143

connects to several conduction tubes that go straight up the plant. Each bud develops a mass of fibrous root branches underground. In this way all of the conduction tubes are connected to a root system.

This relationship between the upper roots and the vertical conductive tubes going up the corn plant is important where phosphorous nutrition is concerned. Phosphorous has a characteristic of being unable to cross over from one conductive tube to another as it moves up the plant.

A situation could arise where this inability of phosphorous to cross over could reduce yields. We will assume you have cash rented on a year-to-year basis a field that is extremely low in phosphorous. You wisely decide you cannot afford to build the soil levels and, instead, apply all of the phosphorous in a starter band on one side of the row. At harvest you may notice that the ears emerging on one side of the plants are excellent, but those emerging on the opposite side are very small. Further investigation would show that the large ears emerged on the side of the stalk where starter fertilizer was applied. Under these conditions it would pay to sidedress part of the phosphorous at cultivation on the opposite side from the starter. Of course, you must remember on which side the starter was placed!

Other nutrient relationships should be considered whenever fertilizers are banded. Whenever possible nitrogen should not be used by itself, particularly in the starter band. **We strongly believe the triad of nitrogen, phosphorous and zinc will always function better in combination than when used singularly.**

A pint of zinc chelate has been recommended for use in the row support because zinc improves phosphorous uptake. A readily available source of zinc in combination with phosphorous will help assure adequate phosphorous uptake during the early growth period when soil conditions are less than ideal for roots and nutrient release. Phosphorous uptake is essential for early, rapid root development.

It is also necessary in the formation of adenosine, di- or triphosphates (ADP and ATP) within the cells. These are responsible for most of the energy changes in the life processes of plants and animals. In this respect zinc may relate to energy release and plant vitality because of its indirect relationship to the formation of ADP and ATP along with its direct relationship to phosphorous uptake and root proliferation. Zinc also functions as the catalyst in many enzyme reactions.

We also believe that nitrogen always functions best in the presence of phosphorous. Since phosphorous is necessary for energy release within the cells its adequate uptake during periods of rapid vegetative growth is most important.

Early research indicates that the nitrogen-to-phosphorous ratio has a bearing on the rate of phosphorous uptake by plants such as corn. We do not believe this ratio is as important as previously thought. There are simply too many soil factors affecting phosphorous availability, nitrogen release and root function to permit the establishment of exact nitrogen-to-phosphorous ratios. A given nitrogen-to-phosphorous ratio must necessarily assume that both elements will remain one hundred percent available within the soil. We know this is nearly impossible because soil conditions can change phosphorous availability within hours, or even minutes, after application.

If it is given a reasonable balance of all nutrients the plant will manage its own ratios of all nutrients. Our primary concern is to prevent "luxury feeding" of nitrogen which could force excessive vegetative growth causing a multitude of deficiencies. While the combination of nitrogen, phosphorous and zinc is particularly effective this does not minimize the need for the other elements.

When fertilizers are placed as shown in Figure 142 on page 231, the normal emphasis will be on nitrogen and phosphorous. An all liquid system gives the opportunity to insert small rates of highly available liquid zinc chelates in the different locations. One pint of an 8% or 10% zinc chelate per acre is automatically included in the row support, plus one pint in either the starter band or sidedress band, for a total of two pints per acre for all crops even though soil zinc levels may be adequate.

If soil zinc levels are low, up to two quarts of an 8% or 10% zinc chelate per acre could be used to take care of the crop on a year-to-year basis while building the soil levels. It is preferable to build soil levels of zinc with low-cost zinc sulphate and use the zinc chelates as outlined to guarantee availability early in the season. Using two quarts, the row support level would be kept at one pint per acre and the other three pints applied in the starter or in combination with the sidedress. If necessary as much as one quart of eight or ten percent zinc chelate could be used in the row support.

If using a dry blended starter the liquid zinc chelate could be sprayed into the dry fertilizer during the blending operation. As a chelate the zinc will remain available even though it is coated onto phosphate granules.

Zinc sulphate can also be blended into a dry starter at three to five pounds per acre for use on row crops. When necessary zinc needs can be totally managed by blending five pounds of 36% zinc sulphate into dry starters for row crops. This, along with the zinc chelate in the liquid row support, is quite adequate for management of zinc.

There are, of course, several combinations for applying nutrients. (As shown in Figure 142 on Page 231).

First, the **row support** system should become a standard practice. The twenty pounds of P_2O_5 as polyphosphate, applied in the row for corn, may be enough if the soil P2 levels are over eighty PPM or where heavy rates of manure are being used. There is no need to maintain soil P2 levels of phosphorous exceeding eighty PPM. Excessive phosphorous can reduce zinc availability. **Row support** permits tapping this phosphorous reserve without sacrificing anything else on the start. Sometimes it may require several years to drain off the excessive inventory to a level where only maintenance is necessary. In that case phosphate would not be included in either the starter or sidedress placements. Straight nitrogen, plus the zinc, could be used for starter or sidedress. Nitrogen can be applied entirely in the starter band, all of it sidedressed at cultivation, or applications split between the two placements.

The next logical question is, how much nitrogen will be needed to produce a maximum yield. Determination of exact nitrogen needs is most difficult. The goal for nitrogen management of any crop is to use minimum rates without placing limitations on potential yield. To be safe we normally recommend an excess of nitrogen. **The key is to keep the excess to an absolute minimum.**

Nitrogen management is not an exact science and there are several key factors relating to its usage that must be fully understood. An awareness of the following nitrogen factors should be convincing enough for many farmers to change their application methods which will ultimately benefit the crops, soils and pocketbook.

1. For many crops nitrogen is the most expensive of the fertility inputs. Therefore, proper nitrogen management offers the best opportunity for reducing overall fertility costs.

2. Nitrogen is the growth element and its intake must be carefully balanced with all of the other nutrients. Excessive nitrogen, particularly when applied too early, can create a host of problems that rob yields. Too much nitrogen in relationship to other nutrients weakens growth, increases potential stress and opens the door for insects and diseases. It can stimulate rapid growth during a time when the soil release of other nutrients is at its lowest level. This alone can cause severe plant deficiencies of other essential nutrients.

3. Preplant, broadcast nitrogen applications may result in nitrogen losses exceeding fifty percent. Losses result from nitrate leaching, gasification and organic lockup. Bathing soils with excessive nitrogen results in the destruction of humus and a deterioration in the physical qualities of soils.

4. All nutrients are important, but calcium and sulfur are especially important for proper nitrogen uptake and utilization by the plant. Calcium is essential for nitrate uptake and sulfur is necessary for the conversion of nitrates into amino acids. Sulfur is helpful for minimizing the accumulation of non-protein nitrogen in plant tissues.

5. Nitrogen recommendations call for the consideration of many factors. Total nitrogen needs will depend upon the following considerations:

A. The nutrient and water-holding capacity of a soil, which may limit production regardless of nitrogen rates used.

B. Levels of humus and organic matter. An accumulation of undecomposed residue may consume considerable amounts of nitrogen while decaying. Humus levels can also influence nitrogen release and retention.

C. Rotation and the preceding crop. Soybeans, for example, may leave thirty to forty pounds per acre of residual nitrogen carryover for the next crop.

D. Manure management as it relates to rates, chemical analysis, method and time of application.

E. Method of placement and timing of nitrogen applications. These two areas are significant; placement within the "zone of updraft" is far more efficient than broadcasting. Timing of the application nearer peak crop needs also influences rates needed.

F. Tillage methods as they relate to residue management or placement. Properly positioned residue, such as cornstalks, will tie up much of the broadcast nitrogen during the decaying process. This must be considered since a nitrogen deficiency could occur if the decaying process coincides with the period of peak crop needs.

G. Fall or spring tillage in reference to its influence on the decay cycle. Fall tillage can start the decaying process and set the stage for completion prior to peak crop needs. Spring management may delay decay and allow competition for nitrogen between the plant and decomposing organisms.

H. Drainage and other factors as they relate to potential yields.

I. Irrigation.

J. Personal experiences on a particular farm. This may be the most important consideration of all, providing the experience is the result of a planned nitrogen testing program. Fine-tuning of nitrogen needs is an essential element of an overall fertility program. Because the determination of correct nitrogen rates for a particular farm is not a simple assignment nor an exact science, knowledge and experience are very important factors in deciding nitrogen inputs.

When nitrogen is broadcast many of the factors mentioned above come into play to influence its fate. As a result, broadcasting requires the use of large rates of nitrogen as insurance that there will be enough left over for the crop.

The form of nitrogen to be used must also be considered. Nitrogen, placed within the "zone of updraft", will be either liquid or dry. We prefer a liquid system, but often work with a combination of liquid and dry starter or sidedress systems. With a liquid system the preferred blending is 10-34-0 polyphosphate with 28 or 32-0-0 liquid nitrogen for nitrogen and phosphate needs.

When using a dry starter, as much as two hundred twenty-five pounds per acre of 45-0-0 has been applied through the planter and placed five inches to the side of the row. This, along with row support, supplied a total of one hundred seven pounds per acre of actual nitrogen. This shows there are very few limitations on rates or forms of nitrogen that can be used, providing it is set out three to five inches to the side of the row. Row support makes it possible to set the starter band out three to five inches without sacrificing the early growth needs of the crop.

Such dry products as 18-46-0 (DAP), 45-0-0 urea, 0-0-60, 0-0-52 and trace elements may be blended to make up any combination necessary for your conditions. When placing dry fertilizer three to five inches to the side of the row we recommend the use of lower cost DAP (18-46-0) rather than MAP (13-52-0) as a source of phosphorous. Our usual concern with the use of DAP is its ammonium (di-ammonium phosphate) when placed close to the seed.

A dry nitrogen form may also be applied as a sidedress. There are several commercial dry applicators that can be mounted on a cultivator and ground driven. Figure 145 shows a modified corn planter being used to sidedress corn with a dry source of nitrogen.

Placement is the absolute key for reduction of nitrogen costs. This system prevents the overexposure of nitrogen to the entire soil system and locks the mobile nitrate into the "zone of updraft" thereby eliminating many of the uncontrollable factors that make nitrogen recommendations so difficult.

Correct placement removes most of the guesswork involved in the determination of application rates and builds predictability and precision. With proper placement and approximately three years of field tests you will be able to determine optimal nitrogen levels needed

Fig. 145

to produce a maximum yield under your own conditions. In the final analysis most farmers are amazed at how much nitrogen they have been wasting in the past.

Your first task is to determine your own method of placement within the "zone of updraft," as illustrated in Figure 142. The options are to place all of the nitrogen through the planter, place it all at the time of cultivation, or to divide the nitrogen between the two positions. We must stress that, regardless of which placement combination you select, it will not change total nitrogen needs if it is all placed within the "zone of updraft."

However, we should be aware that if the election is made to apply all of the nitrogen during cultivation we must assume all of the risks of weather and other factors that could prevent cultivation. Soil types, drainage and other factors will ultimately determine whether one can risk application of all nitrogen during cultivation.

You are now ready to select the application rates for a particular crop you may be growing, such as corn, milo, sunflowers, cotton, etc. For practical purposes let us assume the crop is corn, with the same principles of placement, however, applying to the other crops as well. The only variable will be the nitrogen application rates.

We use two general rules to determine the total nitrogen needs for a particular crop.

Rule 1: Because we can experience losses which range from forty to sixty percent with broadcast nitrogen we suggest that the total nitrogen be reduced by forty percent if it will be banded in the zone of updraft instead of broadcast. As an example, if you have been broadcasting one hundred sixty pounds of actual nitrogen per acre, you should be able to produce as well with about ninety-five pounds of nitrogen per acre that is banded within the "zone of updraft."

Rule 2: When nitrogen is applied pre-plant, as fertilizer or manure, you can assume your crop will recover and utilize fifty percent of this nitrogen.

Under some circumstances part of the nitrogen may be broadcast. Let us assume that we have broadcast a total of forty pounds of actual nitrogen per acre from an application of manure, or possibly from 21-0-0 ammonium sulphate being used as a source of sulfur. Let us also assume that we need a total of ninety-five pounds of nitrogen per acre placed within the zone of updraft, with no broadcast, to produce a maxium crop of corn. Since only about fifty percent of the forty pounds of nitrogen broadcast would be recovered we would reduce the application within the zone of updraft by twenty pounds, or from ninety-five pounds to seventy-five pounds of nitrogen per acre. The fifty percent figure is an educated guess and may actually be more or even less. However, we believe it is a good figure to use to keep nitrogen on the safe side, particularly when we are depending upon nitrogen from manure sources.

For tests of the nitrogen rates for either a corn-soybean or a continuous corn rotation we suggest beginning with a base of one hundred pounds of "usable" nitrogen per acre. This would include fifty percent of any nitrogen broadcast and one hundred percent of the nitrogen placed within the "zone of updraft." The "zone of updraft" would include nitrogen positioned as row support, starter and sidedress. We recommend varying the base rate by ten pounds of nitrogen per acre, going as low as seventy pounds, and possibly as high as one hundred thirty pounds per acre.

You must also determine the normal rate to use on the balance of your acreage that is not included in the test sites. Cutting the nitrogen rates by up to forty percent is quite a drastic change from the norm. This may cause undue concern on the part of your banker or landlord! You must use a rate that you can be comfortable with on the bulk of your acreage while you are in the nitrogen-testing phaze.

However, one hundred forty to more than one hundred sixty bushels of no. 2 corn is presently being produced on a wide variety of soils in a broad geographical area using a total of ninety-five to one hundred pounds of nitrogen that has been placed in the zone of updraft. This is being done on a corn-soybean, or continuous corn rotation without the benefit of manure, irrigation or cover crops. Your confidence will grow with knowledge and experience.

Altering your rates will require time and precision in the planning, adjustment of equipment, marking the sites and measuring results. Your sole purpose is to test only one variable — nitrogen rates. The ease of testing for nitrogen rates will depend mainly on three factors:

—Your method, or methods, of application of nitrogen,

—Your knowledge and experience in adjusting equipment application rates, and

—Whether you are using straight nitrogen or nitrogen in combination with nutrients such as phosphorous or zinc.

The simplest testing procedure would be the election to apply part of the nitrogen through the planter and the balance while cultivating. This simplifies matters because a base level could be applied at planting with other ingredients needed, such as phosphorous and zinc, and then straight nitrogen could be used in the cultivator application. It would be easier to vary the nitrogen rates during cultivation if no other ingredients were added at the same time.

Regardless of your placement system you must accurately vary the nitrogen rates to determine optimal levels of nitrogen needed under your soil conditions. In setting up meaningful test plots the following guidelines should be considered:

—Nitrogen must be the only fertility variable in your test plots. The row support nutrient, additional phosphorous, zinc, etc., blended with the nitrogen, must remain constant on all plots. Obviously, it would be simpler if straight nitrogen was applied. The goal is to have nitrogen the only variable within a complete fertility program.

—A soil that is as uniform as possible should be selected for each plot. If your soils are of two major types the test plots should be set up on both types. If you are on one of those farms where soil types change every one hundred feet plots should be selected that are as nearly similar as possible.

—Test sites should be of adequate size to obtain significant results, probably a minimum of five acres.

—Hybrid testing is also an important factor. For example, all corn numbers will not perform equally at a given nitrogen level. These traits should be investigated with your seed company and at least two of your major seed numbers should be included in each plot.

—All other factors should be kept as nearly uniform as possible across all plots.

Our goal is to fine-tune application rates of nitrogen over a period of time so we may be assured we are getting maximum return on our total nitrogen investment. We must stress, though, that supplemental nitrogen requirements may gradually decline over a period of time with a well managed program that considers the chemical, physical and biological needs of the soil. As you grow in knowledge and experience your overall management becomes more precise and consistent, leading to a continuous improvement of the entire soil environment.

A totally managed, biologically active soil can release considerable amounts of soil nutrients, including nitrogen. We are unable to predict the level of atmospheric nitrogen fixation by soil organisms. We do know that it can be rather significant in well managed soils. This windfall can be capitalized on by periodically re-evaluating nitrogen needs with the use of test plots. After a few years we could make the pleasant discovery that our nitrogen needs can be reduced by another ten to fifteen pounds per acre.

As our soils become better balanced we may also find it profitable to increase nitrogen rates. We must think in terms of finding the most profitable level rather than merely in terms of a reduction of rates. There should be no objection to increasing an input as long as it shows a profit. Nitrogen testing must be an ongoing part of an overall fertility program.

To increase production we must indentify and remove as many yield-limiting factors as possible. A major concern in testing a single input such as nitrogen is to minimize false readings. Once a maximum nitrogen level has been settled on for individual conditions we must be reasonably certain that a deficiency of another element is not preventing an additional increment of nitrogen from expressing further yield increases.

The zone of updraft (Figure 142) fertility placement system is both effective and efficient. The following statements provide a recap and overall review of this system:

—Because of the differences in nutrient accessibility and recovery banding is a more efficient method of providing annual fertility needs than broadcasting.

—The **row support** nutrient package may be considered the most important aspect of the placement system because it provides for the start and gives important options in the starter and sidedress positions.

—Through proper placement the savings in nitrogen go beyond the monetary savings of using less nitrogen. There are many direct and indirect benefits to the soil, environment and plants by the simple elimination of excessive nitrogen. Nitrogen contamination of ground water could probably be eliminated with proper nitrogen placement.

—If you are using an all-liquid system we recommend using a polyphosphate in the starter band for the same reasons outlined for its use in row support.

—There are definite advantages in selection of a liquid system instead of dry. Since the row support must be liquid, a liquid system eliminates setting up two separate handling systems. The blending of liquid nitrogen, phosphates, and trace element chelates is simple, very accurate and can be done at the farm level. Liquid provides the most thorough method of blending ingredients.

—The "zone of updraft" is an extremely efficient place to position fertility because of the presence of roots and the influence of these roots on soil aeration and the movement of water and nutrients. This makes the "zone of updraft" the zone of maximum recovery.

As we learn more about the soil-plant system it becomes obvious that placement is the key to improving fertility efficiency. Efficiency is achieved by producing maximum yields with minimum per unit cost. Since there are no controls on production within the industry our only hope for survival as an individual farming unit is to be considerably more efficient than the average.

FERTILITY MANAGEMENT FOR NO-TILL

Managing a successful no-till operation is considerably more complex than planting under more conventional tillage systems. **No-till's success or failure will depend on the ability to manage soil temperature, soil aeration and soil fertility.** In our experience the vast majority of no-till failures have resulted from the failure to recognize and manage these three areas.

Reduction of production costs is an important consideration, but in the process we must also raise and hold yield levels at higher plateaus. In the attempt to cut costs and simplify our operations we must not lose sight of the need to manage the four essential basics of **soil aeration, water, residue decay** and **fertility.**

No-till leaves no options in the control of residue since it is all left on the surface as a necessity for minimizing water runoff and erosion. In comparison with conventional tillage methods water insoak is normally improved with no-till. However, water retention and residue cover can seriously impede soil warmup in the spring. Using no-till, we tend to plant in colder soils. The soil temperature factor is an important consideration in the management of no-till fertility.

Medium to heavier-textured soils tend to become increasingly more compacted with no-till. The decay of residue on the surface results in the lowering of internal humus and a gradual deterioration of the soil's physical qualities.

By themselves cold soils and poor aeration are enough to cause serious plant deficiencies, regardless of soil fertility levels. By now we are well aware of the roles temperature and soil aeration play in nutrient release, availability and root function. A primary concern with no-till is to assure adequate nutrient recovery by the plant. Ultimately, this will determine our success or failure with the practice.

A successful fertility system for no-till must consider the proper placement of fertility and also overcome the soil temperature and aeration factors. To this extent we must be aware of two basic principles that relate in particular to no-till.

First, the no-till farmer must visualize the concept shown in Figure 146, that he is basically farming within a box that measures fourteen to sixteen inches wide and five to seven inches deep. This is the "zone of action." On medium to heavier-textured soils most of the crop's roots will be found within this box or volume of soil. As a result fertility and aeration must be properly managed within this block of soil.

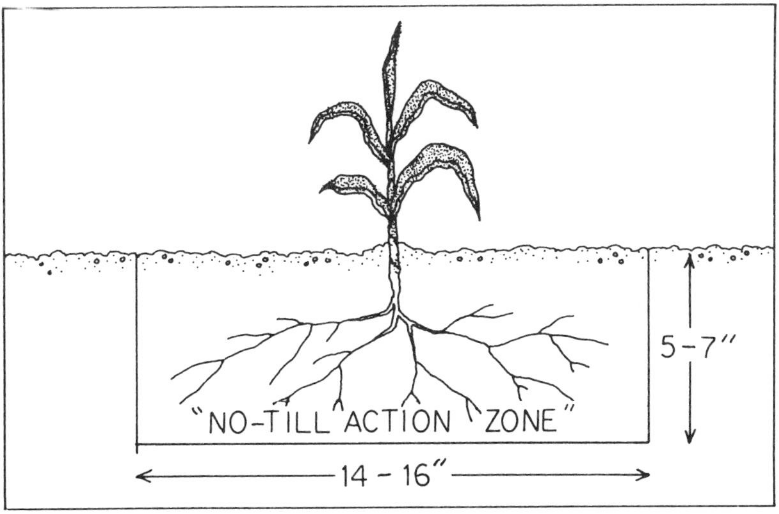

Fig. 146

Second, as shown in Figure 147, an attempt to manage no-till fertility, especially on medium to heavier-textured soils, by braodcasting or dribbling onto the surface, is sheer folly.

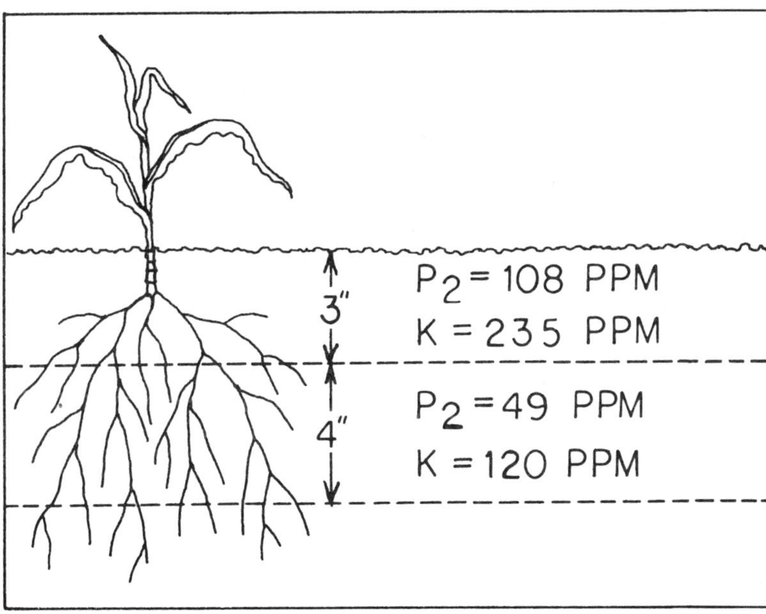

Fig. 147

Note the extreme differences in phosphorous and potassium levels between the top three inches and the lower four inches of soil. We have often observed no-till failures because the lower four inches of soil was actually low in fertility even though there was an abundance above. The top two to three inches of soil is prone to drying out and as a result it is simply too unpredictable to count on fertility coming from this zone.

To overcome the soil temperature factor and to properly manage the areas of soil aeration and fertility, we have designed the "5 x 5 Soil Fertility — Soil Aeration System" shown in Figure 148.

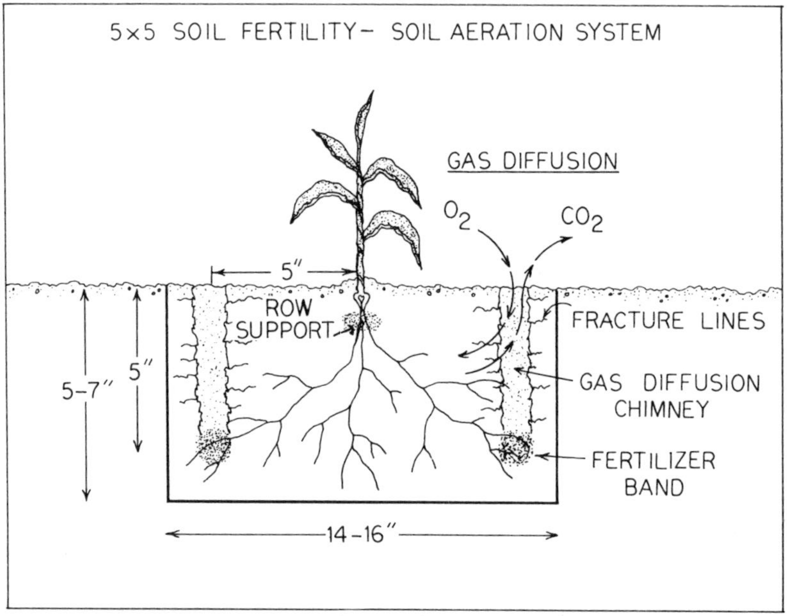

Fig. 148

A **Row Support** nutrient is applied directly into the row with the seed and the balance of the fertility needs placed in a band on each side, five inches out from the row and five inches deep.

With no-till the seed is usually placed in a less than ideal seedbed. Nutrient availability is greatly reduced in cool or wet soils. Under such adverse conditions root growth and function is greatly impaired. The function of **row support** is to assure nutrient availability for the developing plant.

The **row support** concept has been explained earlier in this unit and recommendations for various crops given in Figure 126. For no-till planting the **row support** system is a "must."

The 5 × 5 configuration of placing fertility needs five inches out from the row and five inches deep is significant. The five-inch depth is accomplished by using a rolling coulter followed by a narrow penetrating blade. This arrangement can be fabricated on the farm and there are also several good commercial systems available, such as the Bowman shown in Figure 149. By placing the opener five inches to the side high rates of fertility can be used without concern about interference with germination. This distance also prevents the shank from disturbing the soil around the seed.

The five-inch depth will nearly guarantee adequate moisture around the fertilizer band for root function and nutrient recovery during the full season. In semi-arid, non-irrigated sections of the country it would be a good idea to place it one or two inches deeper.

The five-inch slots serve another important function. They serve as aeration chimneys for the diffusion of oxygen and carbon dioxide. Carbon dioxide diffuses outward while oxygen diffuses into the soil. The lateral diffusion of gases from the root mass through the slots or chimneys is more effective than when it has to come through the surface. Soil oxygen is essential for root growth and nutrient uptake. In addition to providing fertility depth and soil aeration the slots also increase water absorption into the root mass.

An ideal setup would be to use row support on the seed, a liquid fertilizer on one side of the row and a dry fertilizer on the opposite side. This may seem complicated, but it would assure total management of the two main areas that cause no-till failures — soil aeration and soil fertility.

Dry fertilizer would not necessarily have to be applied every year. Potassium, sulfur and trace elements are most economically managed with dry fertilizers. The soil levels should be properly built with all nutrients, including limestone, before entering into a no-till program. Another possibility is that these nutrients could be managed by reverting to conventional tillage every fourth or fifth year if it would not present too serious an erosion hazard.

Combinations of dry fertilizers could be banded every second or third year while planting, or possibly as a separate operation. If potash, sulfur and trace element combinations are banded it may be wise to alternate the application from one side of the row to the other. This would be particularly important if everything was applied on an annual basis to prevent the possible buildup of salts and oversaturation of nutrients. Trace element applications, particularly copper and boron, must be managed very carefully to prevent toxicity problems from developing over a period of time.

If the election is made to apply dry materials every second or third year it may be feasable to utilize an old corn planter chassis with the

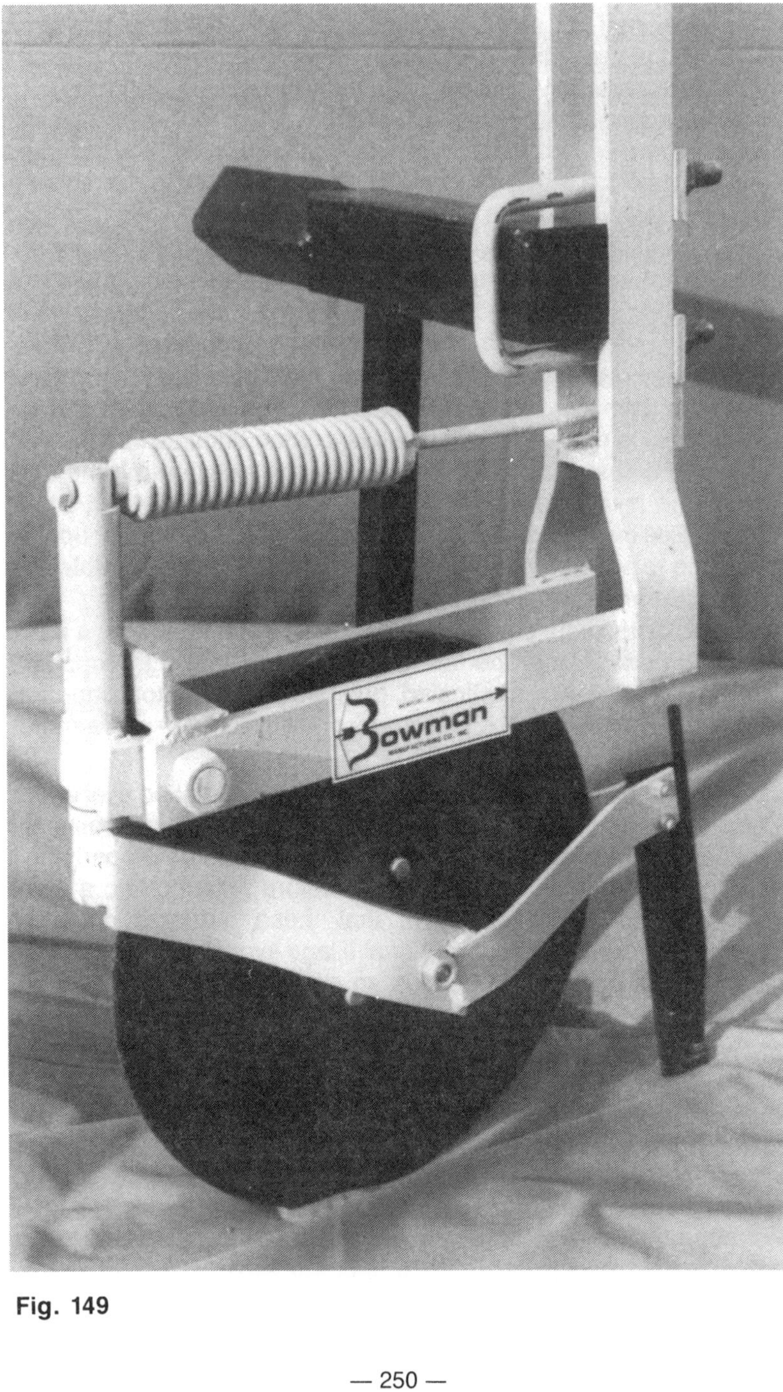

Fig. 149

dry attachments, as shown in Figure 145, to be used as a separate operation. By making a periodic, separate operation the power utilized for planting could be considerably reduced in comparison to the double application. We believe the five-inch depth is important, so it definitely becomes a power factor. A liquid application system on the planter would be simple to use if the decision were made to apply on one side only while planting.

If you decide to apply a dry band in a separate operation and another band while planting it would simplify logistics and handling of materials to set the planter up to use liquid to match the row support. On the liquid side, a combination of 10-34-0 polyphosphate and liquid 28 or 32-0-0 nitrogen, plus the liquid chelates, could be blended to supply the balance of the nitrogen and phosphorous requirements while planting.

Exact fertility recommendations depend upon soil fertility levels and the crop to be grown. Our purpose is to outline no-till principles and systems rather than to attempt precise fertility recommendations.

No-till can be successful providing the soil basics of aeration and fertility are adequately managed. Everything necessary must be done for their management. Once the problems are understood and the systems set up to manage them success will be the rule and not the exception. A total program must be built around the soil and crop's needs rather than for convenience only.

CROP SALVAGE PROGRAMS

Farm profits are determined by the interaction of yields, prices and production costs. If a farmer is to prosper and stay in business he must do everything possible to assure maximum yields with minimal production costs.

Crop production involves the interaction of many factors, some of which are beyond one's control. Farmers must maximize control of those factors that are controllable and minimize the affect of adverse factors beyond their control. To this extent we must think in terms of removing crop limiting factors rather than the application of unrelated practices and inputs in the hope of increasing yields. All production inputs must be designed to eliminate or prevent problems that could arise.

But, even with the best management problems can develop and crops may fail. Failure can range from break-even yields to total loss. However, crops never fail without a reason, and when a crop fails we must carefully appraise the situation to determine precisely the causal factors leading to the failure. We can not simply blame the weather if we are at fault by not initiating practices that could eliminate or minimize weather hazards.

Failure can be rare providing we have the capability to recognize and plan an overall program around the removal of crop-limiting factors. By using this approach and also being prepared to initiate a salvage program if needed we should seldom experience failure.

To be successful a farmer must be aware of the factors which lead to crop failure and we have catalogued these factors under three major groupings, as follows:

GROUP I: The Human Factor. Man-related problems brought about through mismanagement, errors and carelessness. The causes of crop failure within this category are the easiest to control or prevent. They include such areas as:

—Selection of poorly adapted seed varieties,
—Poor timing,
—Planting too shallow allowing the soil to dry out,
—Excessive root pruning while cultivating,
—Overworking the seedbed causing excessive compaction,
—Over or under-planting,
—Poor maintenance of equipment with untimely failures,
—Plugged starter fertilizer outlets on one or more rows,
—Failure to control weeds and mismanagement of herbicides,
—Mis-use of nitrogen,
—Poor fertility spreading patterns,
—Convenience and hurry!

GROUP II: Weather or season-related conditions that are partially or entirely under a farmer's control, such as:

—Poor drainage that can be corrected with tile or ditches.

—Mis-management of residue, making fields more prone to surface-sealing, runoff and competition from decay,

—Plow or disc pans that will restrict root growth and water management increasing the risk of crop failure,

—Tillage practices that increase rather than decrease weather-related problems,

—Failure to cover such natural disasters as hail with insurance.

GROUP III: Factors totally beyond our control, such as:

—Killing frosts that arrive too late in the spring or too early in the fall,

—Severe flooding, heat or drought.

It becomes quite obvious that we have some control over nearly every factor relating to crop failure. Our major concern is that we do not inadvertantly, through our own mistakes, set the stage for failure. Another concern is that we recognize developing problems in time to take corrective action before the damage becomes terminal or beyond correction.

Too often problems are recognized too late to overcome them. How comforting it is to drive around our fields and observe the apparent abundance as reflected by the end rows. How devastating it can be to discover disaster beyond the end rows while harvesting! What seems most tragic is to discover large areas of the inner field with losses that could have been salvaged if discovered in time.

Salvage success requires a two-phased program. First, the problem must be recognized in its earliest stage. Next, we must be prepared to initiate corrective treatment. Proper diagnosis and treatment are the signs of a good doctor. As farmers we must become good plant doctors because our economic health depends upon it.

In a well managed production program the problems that develop are usually weather-related. Mother Nature has the capability of putting the best planned program into a situation needing salvage. Often the salvage situations involve the need for soil oxygen and additional nitrogen. Compacted or water-logged soils become depleted in oxygen which slows or stops root growth and nutrient uptake, causing a combination of plant deficiencies. When air is introduced back into the soil plants will begin to recover if there is still sufficient nitrogen available. Most nutrients, with the exception of nitrogen and possibly sulfur, remain in place and are not lost by the lack of soil oxygen. However, anerobic soil conditions can cause severe and permanent losses of nitrogen through the process of denitrification.

Denitrification losses usually occur on heavier-textured soils. This process is the result of anerobic soil organisms removing oxygen from the nitrate molecule and changing it back into a gas which can escape from the soil and return to the atmosphere. On light-textured soils too much rainfall can cause nitrate loss through leaching. Nitrogen losses, whether from denitrification or leaching, must be replenished to prevent crop losses. Without nitrogen and soil oxygen crop failure is guaranteed.

Nitrogen losses can be replaced in several different ways depending upon the situation. Cultivation is an ideal method of replacing nitrogen and improving soil aeration at the same time. There is no better salvage tool than the cultivator. In reality cultivation can always be considered a salvage operation. We cultivate to improve soil aeration, destroy weeds and to apply a nutrient. Accomplishing any of these goals will improve crop potential and in this respect could be considered salvage. However, we usually use salvage programs as a means of getting a crop out of serious trouble.

Cultivation is an ideal method providing it can be done before the crop grows too tall. There seems to be a need for the development of equipment similar to detasseling machines that would have the capability of light cultivation and application of fertility or foliar at any time during the season regardless of plant height.

The cultivator is an ideal way to spot-treat fields in a fine-tuning process. Figure 150 illustrates a field situation that may be benefited from spot treatment during cultivation. As shown in Figure 150A a field may have several such areas. In this case the light, sandy areas

Fig. 150

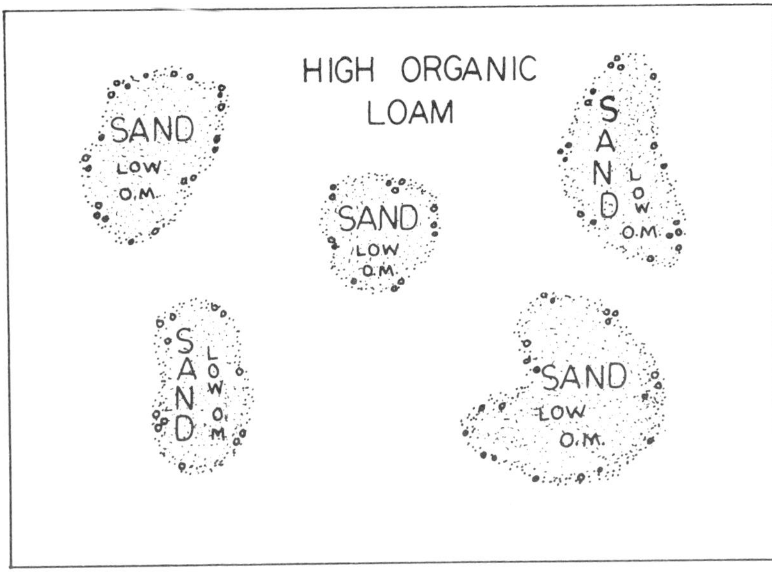

Fig. 150A

can profit from an extra twenty to thirty pounds of nitrogen because of a combination of low organic matter and excessive leaching. The additional nitrogen may prevent unnecessary yield losses (Figure 151).

Fig. 151

A cultivator setup with an electric pump and controlled by an on-off switch, as shown on Page 224, would provide an ideal way to salvage this area.

Cultivator application (Figures 152, 153 and 154) is ideal for salvage or under normal conditions to be used as part of an efficient nitrogen management system. Notice that in all instances the farmers are placing their nitrogen very close to the row to keep it within the "zone of updraft." The center between the rows is probably the poorest place to position any nutrient because of wheel track compaction and its distance from the root mass.

Problems with nitrogen deficiencies may appear when it is too late to cultivate. If this happens, as on the field in Figure 151, it will not be noticed by observation of the end rows. To locate these problem areas you must walk into the different soil areas or fly over the field two or three times during the year. Both methods are highly recommended.

If you fly over your fields carry a farm map so you can record problem areas as you see them. The key is to spot these areas before permanent damage sets in. Corrective action should be taken as soon as you observe a change in color. In the later growth stages you may have to resort to an aerial application of nitrogen or use of special equipment designed to go through tall corn.

For aerial applications we recommend using 45-0-0 dry form urea. Seventy-five to one hundred pounds of the 45-0-0 material will provide thirty-three to forty-five pounds of actual N per acre. At this rate the urea will not burn the leaves. The granules that become lodged in the leaf whorls will take on moisture and actually foliar feed nitrogen into the corn plant. Aerial applications of dry urea can be very effective.

In addition to sidedressing with cultivation and applications by air you will find that under certain conditions foliar feeding of nitrogen can be an important means of salvage.

Nitrogen deficiency causes a loss of chlorophyll resulting in discoloration and reduced sugar production. Since roots do not produce sugar they must depend upon a continuous supply from above to grow and function. By slowing sugar production a deficiency of nitrogen can cause root failure.

Corn or other crops should not be permitted to remain yellow for extended periods of time. After seven to ten days, if conditions do not improve, we recommend an application of five to six pounds of actual N per acre using liquified urea as the nitrogen source, plus one pint of an eight to ten percent zinc chelate. Be certain to use urea because it is the only nitrogen source that can be effectively foliar fed. The 45-0-0 urea nitrogen source should be combined with

Fig. 152

Fig. 153

Fig. 154

ten to fifteen gallons of water for ground equipment application and applied directly over the row. The nitrogen will increase chlorophyll and sugar production, which in turn will provide the energy for root growth and function. Zinc is important in the salvage process.

The success of a salvage program depends upon correct diagnosis and prompt treatment. Factors other than soil aeration and the lack of nitrogen may also cause a need for salvage. Adverse weather conditions may negate weed and insect control and require salvage programs to save the crop. We have concentrated on the correction of soil aeration and nitrogen deficiencies only because these are the two most prevalent conditions we are likely to encounter.

Certain other deficiencies mimic each other so closely in appearance that they may confuse the most seasoned farmer. Several different conditions can result in chlorosis or a loss of chlorophyll. Obviously, a loss of chlorophyll will reduce photosynthesis or sugar production and lead to serious setbacks for both the plant and the farmer.

Figures 155 and 156 are schematic drawings to illustrate the many ways in which photosynthesis may be reduced. They show that magnesium is an integral part of each chloraplast.

The daylight phase (Figure 155) shows that iron, sulfur, nitrogen and manganese are also necessary for the structure and function of the chloroplast. A deficiency of any of these five items can cause a loss of chlorophyll and result in a yellowing of the plant. If chlorosis appears it simply means that the plant has lost its ability to utilize the energy of the sun to convert carbon dioxide and water into sugar. A loss of sunlight will also cause chlorosis.

During daylight hours the plant manufactures sugar and at night (Figure 156) the trapdoor opens to permit the chloroplasts to drain their accumulation of sugar into storage, growth, maintenance, etc.

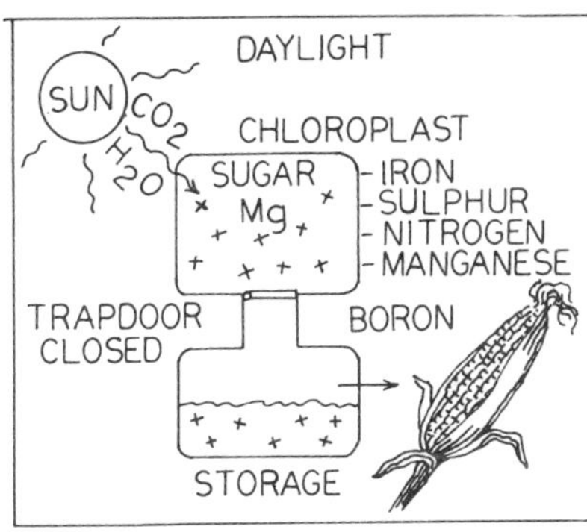

Fig. 155

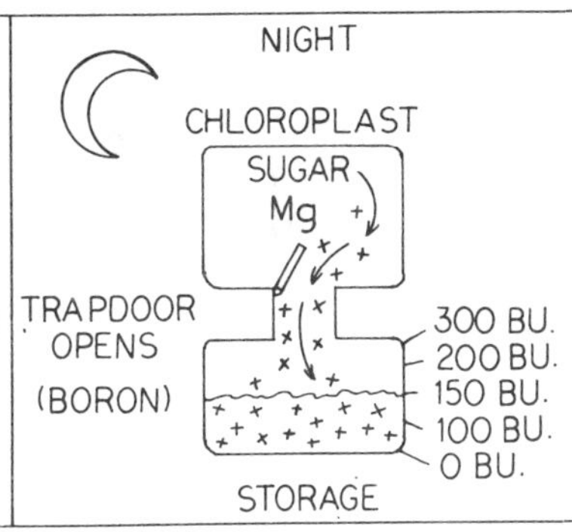

Fig. 156

Plants produce sugar by day and grow at night. The trapdoor must open to permit the exiting of sugar from the chloroplast into storage or production. We show boron as a necessary ingredient to cause the trapdoor to open for the sugars to be transferred. This makes boron essential. If the sugars cannot move out of the chloroplast it tends to build a "diabetic" plant which becomes grossly inefficient. A plant that backlogs sugars is also easy prey for insects. If a plant were perfectly efficient its top yield potential could be amazing.

An interesting instrument to discuss is the refractometer which is used to measure plant efficiency. The refractometer indirectly indicates sugar levels. At present we are using it to check sugar levels in plants at different times of day and under different managment practices. The readings are compared during the early morning hours and also later in the day. High readings late in the day indicate an efficient plant. A low reading in the early morning hours in comparison to late afternoon indicates efficient movement of sugars into growth or storage.

Well managed alfalfa has been found to show a high reading of twenty-six or higher in the afternoon and less than six in the morning. For best quality feed or highest energy we believe the forage should be cut late in the day when its sugar content is the highest. The refractometer is an interesting instrument that can open the door to investigative work at the farm level to not only help increase our knowledge, but also to stimulate the desire to learn and create excitement in our profession.

Obviously, many things can affect the plant resulting in a need to salvage. We use Figures 155 and 156 to emphasize how complex the soil-plant system is.

Our goal is to increase production very simply by removing or minimizing crop-limiting factors. To remove crop limiting factors we must first be able to identify them. This is what education, coupled with experience is all about. In the complex business of farming continuous education is essential and we must each develop the desire and self-discipline to pursue it vigorously. Knowledge and its proper application will stress prevention and minimize the need for salvage programs.

KEYS TO GOOD HEALTH

It seems appropriate to close this book with a discussion of health, something very personal and important to each of us. Regardless of how successful we may be in our business and financial matters we will not be able to fully enjoy our successes without good health.

Because of our intelligence and ability to think and make choices humans are selective and imaginative. We are capable of bringing about changes and are rarely content with the simplicity of nature's ways. For example, we can not perceive letting such processes as natural selection or survival of the fittest regulate our genetic strength. As a result the human species tends to accumulate more genetic weaknesses than would be permitted under nature's unforgiving ways. We attempt to substitute technology for natural laws.

On the other hand, because of instinctive guidance, animals are more predictable and consistent in their survival behavior. They make fewer mistakes in their physical maintenance and seem to instinctively understand how things were meant to be and simply do not divert from the established patterns. If we disregard accidents and predation animals may stand a better chance of living out their lives to their fullest genetic expressions of health and longevity than humans.

Humans are similar to the animals around them but differ in one very important respect. Animals are endowed with the trait called "instinct." Humans appear to have lost many of these survival instincts and have substituted intelligence and what is often called "common sense" in its place. It is through this trait of common sense that we may be permitted a chance to regain a proper perspective of ourselves and life in general.

Because of intelligence and the ability to make choices we have by choice strayed quite far from nature's intended ways. Because of technology we tend to ignore prevention of health problems while placing our faith in scientific remedies. We are at the point where we seem to have developed a *species isolation syndrome* that subconsciously leads us to believe that we are far above other creatures and the natural laws that govern them. However we must realize that we are regulated by the same forces that regulate all other creatures and in this respect must live within the bounds of our resources and natural laws.

Good health requires the understanding of our body and its needs as well as we can and to then simply concentrate on the removal or minimization of as many health limiting factors as possible. In this respect we must recognize the need for "in the ball park" balances and the avoidance of long-term exposures to either excesses or deficiencies.

Balances with the avoidance of excesses or deficiencies must apply to nutrients such as carbohydrates, vitamins and minerals as well as exercise, sleep, or the use of alcohol or tobacco. In most of these categories excesses may be as detrimental to health as deficiencies. While we may "sin" occasionally with respect to our body's needs it must be the exception and not the rule.

In addition to the above generalities the author believes there are four important areas that influence one's physical well-being. These areas are *food and nutrient supplements, water sources, exercise* and *mental attitude.*

The author is not a physician or certified nutritionist and because of the complex nature of this subject presents a "warning label" that very few health authorities can be taken completely at face value. Instead the subject must be approached by the process of exercising your own common sense, which hopefully has replaced lost natural instincts.

FOOD AND NUTRIENT SUPPLEMENTS

Management of our food intake may be the most important area of responsibility. To some extent we are the product of everything we have consumed.

We must first try to view ourselves as we were meant to be before there was modern food technology that included food refinement, food supplementation, additives, preservatives, the vitamin industry and even supermarkets. Let us pretend to go back in time to when our natural instincts dictated the menu, always assuming an adequate selection of natural foods was accessible. We must ask ourselves what our bodies were meant to consume. We are not, as some animals, limited to one source of food, but fortunately can utilize a broad range of foods.

A personal belief is that we are by design meant to select our food as "close" to the soil as possible. Consuming the entire plant is as near as we can get to the soil as a source of nutrients. Fruits are next and followed by seeds. Animal products such as milk and eggs are farther removed from the soil with meat products being the last. In other words we are better equipped to be herbivores than carnivores. This may be construed to mean that we are designed to take the bulk of our diets from whole plants, fruits and seeds. It does not mean that we should completely avoid meat and animal products. The implication is that there must be an intelligent balance between plant and animal sources of food, with whole plants, fruits and seeds being the foundation upon which to build.

An ideal diet should entail the consumption of a variety of plants grown on a well managed soil and supplemented with fruits, seeds and animal products. Balance is the key.

If ingestion of excessive meat and animal products seems to result in cholesterol problems they could be caused by a lack of exercise and deficiencies of other nutrients rather than simply from the use of animal products. Symptoms result from the problem and we must not construe the symptom as the problem. It all reverts to the philosophy of balances and the avoidance of long-term exposure to excesses or deficiencies.

Misinformation, self-interests, even ignorance can interact resulting in an atmosphere of fear and doubt that in turn may lead to overreaction which can cause a whole new set of problems. Over-reacting can bring about dietary changes that may be totally unjustified and even dangerous. One example of this is the fear of cholesterol that has caused a mass movement of a segment of the population away from use of animal fats to vegetable oils. Is it justified or will it result in a whole new set of unforeseen problems? Who is the final authority on such issues? It has recently been discovered that we have both good and bad forms of cholesterol. It is easy to look at single causal factors to find cures. This probably results because the subject of health may be too complex to comprehend.

If we accept the premise that the base of our diets should come from plants, seeds and fruits we must also recognize the need to grow these foods properly and the importance of variety. Our goal with all of our crops is to grow them as genetically normal as possible with a minimum of deficiencies or excesses. This is an important concept, particularly in respect to the whole plant which can vary in its chemistry depending on the conditions under which it grew.

The nutrient levels within a plant will vary much more than those found in the seeds from the same plant. Seeds will vary, but nature makes every effort to stabilize the nutrient balance within the seed.

A plant can even become toxic with excessive nutrients such as iron, aluminum and heavy metals. To assume that a plant has average nutritional values can be very hazardous to one's health. The only way to assure proper balance in a plant is to grow vegetable crops on a biologically active and nutrient-balanced soil. Applications of too much or the wrong fertilizers is a major error many gardeners make.

The indiscriminate fertilization of a garden as shown in Figure 157 is a good way to kill essential soil life and grow junk food. We must run a complete soil analysis and fertilize our gardens accordingly. Because we may be working with only a fraction of an acre it is very

Fig. 157

easy to over-fertilize. The gardener in Figure 157 may be applying fertilizer at ten to twenty times the rate actually needed.

Management of trace elements is very important to your vegetable plants and for your health. If twenty pounds of boron material, for instance, is needed, keep in mind that this represents only about one-half pound of material for each one thousand square feet. A garden grown with variety on a well balanced, biologically active site as shown in Figure 158 will produce the nutritional foundation for superior health. How fortunate those are who have the land and know how to produce and properly preserve a variety of nutritious vegetables that are free of excesses, deficiencies and toxins.

The intake of additional supplements such as vitamins and minerals must also be reviewed. If we look closely at ourselves we must conclude that our bodies were meant to accept their nutrient needs from biological sources. The term biological sources is used to emphasize that almost all of our nutrients should come via other "living systems." Most plants are designed to efficiently convert raw minerals into complex organic systems that we can utilize. Our bodies will fare much better taking their minerals from plants than from raw minerals.

My belief is that plant sources of vitamins and minerals are superior to manufactured ones if for no other reason than that the natural sources carry a biologically synthesized nutrient package that must frustrate the pure chemist. Imagine trying to manufacture and encapsulate all of the nutrients found in a properly grown plant into a pill with expectation that when consumed the pill will perform equal-

Fig. 158

ly as well as the whole plant. It simply will not perform as well because of the effect of the unknown factors in the living systems. These unknown factors are the result of the unique way nature packages her foods.

Supplementation with artificially synthesized nutrients called vitamins has been promoted as a means of protecting ourselves from bad eating habits. This is a dangerous situation that has often led to mega-dosing of vitamins and minerals. The American diet has degenerated to the consumption of highly processed foods fortified with vitamins, minerals, preservatives, artificial coloring and flavoring and whatever else is needed to preserve them. In addition to consuming these we further insult our bodies with excessive salt and carbohydrates. It will take more than a few daily vitamin pills to save our bodies from this onslaught.

It is time to sit back and reflect. Let us again ask ourselves how things were meant to be. Answering this question gives our common sense a chance to come into play and clear the air. We must not be blinded by man's technology. On the other hand, nature's technology has been time-proven and fine-tuned to assure success. We must not lose this perspective.

WATER SOURCES

Water in its purest form contains only hydrogen and oxygen. However water is also a pretty good solvent and will absorb nearly everything with which it comes into contact.

Water can be highly controversial. Because its flavor is dependent upon whatever has been absorbed in it personal preferences in relation to its taste are often the result of life-long exposures. Some people love sulfur-flavored water while others can not abide it.

There are basically three kinds of water to consider in satisfying our body's needs. Biological water is that which is found in living tissues such as in apples or watermelon. A second source is rain water and formerly it was nearly free of minerals. The rain barrel of old provided a source of soft water that far outperformed well water in sudsing action. A third source of water comes from deep under the ground. This source is considered hard because of the many minerals it absorbs as it filters through the soil.

The question is, which sources of water are best suited for our bodies. Again, we must ask ourselves how things are meant to be. Under natural conditions most animals obtain their water from biological food sources with supplemental needs from ponds or streams.

How important is biological water and utilization of softened or demineralized water? Both are important considerations for the following reasons. First, biological water that is found in foods appears to be the most natural source obtainable. It is difficult to believe that so natural a source could be detrimental in any way to the body. This water is part of the complete organic package the body is designed to accept.

Next, the water found in ponds or streams, unless spring fed, is relatively soft and free from excessive minerals in comparison to well water. It appears that our supplemental source of water was meant to come from rain or other relatively mineral-free sources.

Ground water can be extremely high in dissolved minerals which are then ingested into the body with the water. When we consider that plants are designed to use raw elements, but that humans are poorly equipped to do so it would seem unwise to over-indulge in mineral-laden water. Modern man with his highly developed technology is the only animal on earth capable of drilling deep into the earth for his water supply. After consideration of all these factors it appears that biological or distilled water may be the most natural sources of water for humans to consume.

EXERCISE

We must pause and look at our bodies with respect to physical movement. The vast bulk of the human body is made of muscle, tendons, skeleton and supportive circulatory tissues.

In reality we are "hunters and gatherers" who were designed to cover vast distances in pursuit of food. The original hunt and gather society suffered from a multitude of hardships, but rarely from hardening of the arteries or angina pains.

In today's society there is little real need to physically exert ourselves. Often, the less we exercise the less we wish to do so. We know that the failure of any one system within our bodies can cause our demise and are aware of the problems in association with circulation. A lack of muscle use coupled with excessive caloric intake is nearly a guarantee of health problems often beginning with maturity. Circulatory problems are becoming rampant in all age groups.

During the Korean War autopsies on young men killed in battle showed an alarming degree of cholesterol within the heart arteries. The Vietnam War showed the condition to be worsening. Why? Again the answer is complicated. While nutrition is important, nutrition without exercise is little more than one-half the way to good health. Coupling good nutrition with regular exercise is the preventative measure.

Our bodies are designed to move. Movement from the refrigerator to the table or in and out of the car is simply not enough. Exercise must be sensible to be meaningful for the body. My purpose here is not to prescribe any particular exercise regime, but simply to point out that the body **must** be used to be healthy.

MENTAL ATTITUDE

One's mental attitude is most often a self-reflection of life's personal relationships, physical health, financial situation and life experiences in general. In the same way that physical health may be the product of everything we have consumed our mental attitude may be a reflection of our total life experiences. As physical problems may limit physical activity, so may mental attitudes enhance or limit our successes. Our mental attitude reflects how we see ourselves as well as how others view us.

Most people are interdependent with others and few can be islands unto themselves. Interaction with each other is essential and unavoidable. Personality is a great communicator that reflects a body language more easily understood by others than that expressed in words. The mind can reflect feelings of emotion through bodily expressions that are easily read. This makes our mental attitude of prime concern when interacting with others.

We may at first conclude that a positive or negative attitude comes with practice alone. It is easy to be positive when things are going well and to turn totally negative with adversity. Positive thinking is a happy state of mind reflecting optimism while negative thinking reflects a pessimistic view accompanied by real and imagined fears. Given a choice there is no question which frame of mind we would choose.

It is very easy to sort out life's events as being either good or bad. Our goal must be to dilute the bad by turning adversity into something positive and beneficial whenever possible. We learn by doing and if we are "doers" we will make some mistakes. We must minimize mistakes, but when we make them anyway, why not turn them into lessons which will not only benefit us in the future, but also make them easier to live with?

When we experience adversity we should not carry the burden alone. Whether we turn to our religious faith, seek professional help, or simply share our troubles with family and friends we involve personal relationships that can benefit everyone. Often we need to interact with others simply to keep problems in perspective. For instance it is easy to become so burdened with financial or other stresses that it actually destroys our health and breaks up our families. Proper perspective is important. A financial loss can be troublesome, but the loss of health as a result turns it into a devastating event.

It is helpful to remember that nearly every situation that brings on emotional stress is survivable. The degree or level on which we survive depends largely on how well we have mastered being either positive or negative. Common sense and a positive attitude are both obtainable through practice. They are both necessary for getting the most out of life.

ACKNOWLEDGMENTS

Figure 10	Lloyd Hamrick — In Memory
Figure 51	Richard Coughenour — Sheldon Illinois
Figure 62	Harold Schriefer — Landclearing, waterways, ponds and Conservation Structures — Grant Park Illinois
Figure 72	Don Olthoff — Grant Park Illinois
Figure 77	Tom Nelson — Beecher Illinois
Figure 79	Photo by Soil Conservation Service — Richard L. Dickerson — Conservation Agronomist
Figure 80	Don Olthoff, Dennis Wildman and Bernie Olthoff
Figure 87	Gaylord Lottinville and Sons Kyle and Kevin of Sheldon Illinois
Figure 88	Kyle Lottinville
Figure 91A	Chuck Walters — Acres, USA Publication of Kansas City Missouri
Figures 96 and 97	Compliments A & L Great Lakes Agricultural Laboratories, Inc. — Fort Wayne Indiana
Figures 106, 107 and 108	Compliments of Allied Chemical Company
Figure 128A	Ted Hoekstra of DeMotte Indiana
Figure 130A	Dethmers Manufacturing Company — Boyden Iowa
Figure 131	Kyle Lottinville
Figure 132	Richard Peterson of DeMotte Indiana
Figure 133	Dennis Wildman — Field Service Director for Advanced Agriculture, Inc. Eastern Division — Farmdale Ohio

Figure 134 RM Distributors — Vail Iowa

Figure 149 Bowman Manufacturing Company

Figure 152 Ed McDougall of Grove City Pennsylvania

Figure 153 David Peterson of DeMotte Indiana

Figure 158 Susan Bradley Schriefer

Book Illustrations
by
JAN E. McCRACKEN
Normal Illinois

A special note to my wife, Sue, whose encouragement and efforts were most helpful to my completing this book — many thanks!

Don

INDEX

W

Z